# MOISTURE SENSORS IN PROCESS CONTROL

# MOISTURE SENSORS IN PROCESS CONTROL

K. CARR-BRION

*Warren Spring Laboratory, Control Engineering Division,*
*Stevenage, Herts., UK*

ELSEVIER APPLIED SCIENCE PUBLISHERS
LONDON and NEW YORK

ELSEVIER APPLIED SCIENCE PUBLISHERS LTD
Crown House, Linton Road, Barking, Essex IG11 8JU, England

*Sole Distributor in the USA and Canada*
ELSEVIER SCIENCE PUBLISHING CO., INC
52 Vanderbilt Avenue, New York, NY 10017, USA

WITH 16 ILLUSTRATIONS



**British Library Cataloguing in Publication Data**

Carr-Brion, K.
Moisture sensors in process control.
1. Process control 2. Automatic control
I. Title
670.42′7 TS156.8

**Library of Congress Cataloging in Publication Data**

Carr-Brion, K.
Moisture sensors in process control.

Bibliography: p.
Includes index.
1. Chemical process control. 2. Moisture meters.
I. Title.
TP155.75.C36 1986 660.2′81 86-6317

ISBN 1-85166-005-4



Printed in Great Britain by Galliard (Printers) Ltd, Great Yarmouth

# *Preface*

The most intractable barrier to the full implementation of automatic process control is the lack of adequate on-line sensors. In many cases this is caused by poor selection at the design stage, by lack of knowledge of what is available or by inadequate appreciation of the limitations of the type of sensor used. This book seeks to alleviate this problem in the important field of moisture measurement, drawing on the broad industrial experience of Warren Spring Laboratory and the many companies with whom the Laboratory has worked. Some sources of supply are included of sensors which although not necessarily comprehensive constitute an important source of further information. My thanks are due to the support received from individuals within the Laboratory and many industrial companies and to Jenny Willer for converting an ever-changing manuscript into a readable typescript.

K. Carr-Brion

# *Contents*

CHAPTER 1

# *Introduction*

Water is essential for all known forms of life, and is also omnipresent in the Earth's environment. Water molecules are highly polar—that is there is a non-uniform distribution of electric charge between the hydrogen and oxygen atoms. The results of this are many, but one of primary importance in moisture measurement is that water molecules attach themselves tenaciously to surfaces. Hence, the surfaces in any system are normally contaminated with water from the atmosphere and it is a prerequisite for accurate measurement to ensure the surfaces are cleaned or purged beforehand. Further, moisture must not be allowed to leak into the measurement system or otherwise interfere with the determination—saturated air contains about 17 g water in a cubic metre at 20°C[1]. By fortunate chance many water sensors for use with liquids or solids are only marginally affected by gaseous moisture, which greatly simplifies on-line determinations. Another result of the highly polar nature of water molecules is the formation of so-called 'hydrogen bonding'[2]. This means that water in liquids and solids possesses ill-defined structures of great importance in life and related processes. It is said to exist in many 'forms' within a solid or a solid–liquid mixture. The determination of the concentration of these various forms on-line is becoming increasingly important in industrial measurement and has merited separate consideration in this book.

The presence of water at the correct concentration, or indeed its absence, is required in the vast majority of processes, ranging from food manufacture through biotechnology and fine chemical production to minerals separation and cement manufacture. Its determination in raw materials, intermediates and products is one of the major measurement requirements in the processing industries, and as the use of automatic control spreads under the influence of reliable, low-cost and high-powered microelectronic-based control systems, so is the demand for on-line water determination steadily increasing.

A knowledge of water concentration may be required for many reasons. It affects the transport properties of pastes, suspensions and slurries, as well as the ease of flow of powders from hoppers and chutes. Buying water in a raw material wastes money, selling it in a product makes profits—hence its precise control is of value in company profitability. It may be required to ensure the safety or correct completion of a chemical reaction, or for allowing optimum energy use or material throughput in a process such as drying. The measurement can be required in gases, liquids, solids or combinations of these ranging from thick pastes to foams. Hence, on-line water determination is not just one problem but a whole family of problems, the solutions to which require expertise in on-line sampling, materials handling, sample conditioning, surface physics, complex data interpretation and, last but not least, sensing itself.

## 1.1 HISTORICAL BACKGROUND

For thousands of years water content measurement depended on inference—from weighing, mixing known volumes, or the naturally occurring concentrations in raw materials—and subjective judgement—the plasticity of a clay, the feel of a fibre or the cohesiveness of a dried powder held in the hand. It was part of the knowledge base of the craftsman and was as jealously guarded as the many other secrets of each individual trade.

With the coming of the industrial revolution the need to measure moisture[3] broadened, one of the earlier examples being the control of humidity in textile processing, where hair hygrometers, first described by de Saussure in 1783[4], were used to indicate the conditions in the spinning rooms. The increasing interest in meteorology in the 19th Century resulted in the development of dew point and wet and dry bulb hygrometers for more precise moisture measurement. Determination of water in solids and liquids remained generally dependent on laboratory analysis of samples taken from the process or the previously mentioned subjective tests. It was not until the late 1930s that electronic moisture sensors were first described, while the adoption of the wide range of on-line methods described in this book had to wait until the 1950s and early 1960s.

## 1.2 OBJECTIVE AND SCOPE

The objective of this book is to provide a concise practical guide to the on-line determination of moisture for process control. It is not intended as a source for a comprehensive theoretical background for the various techniques of water determination, beyond that deemed necessary to understand the advantages and weaknesses of the individual techniques. The whole topic area of moisture determination in meteorology is excluded, although many of the gaseous sensors described are very suitable for atmospheric moisture determination. Automatic laboratory methods are also excluded except in the rare cases where on-line (that is measurement without human intervention) use is possible.

It should be stressed that sensor manufacturers and many users have an immense amount of practical applications experience. While some of this may be biased or outdated, it is of great value to workers on the plant but too extensive to be included in a volume of this nature.

While the book is written in an order based on the type of material being analysed—whether gas, liquid, slurry, paste, suspension or solid—readers with problems are first directed to Chapter 9—choice of sensor—where it is hoped that they will readily be able to limit the scope of possibly suitable sensors. The next step suggested would be to read more about the chosen sensor in the earlier chapters, then approach the industrial suppliers indicated in Chapter 10. Thus, they will be able to reject unsuitable sensors (or those working on the margins of their performance, which is highly undesirable in process control measurement), choose the most likely, get a basic understanding of their strengths and limitations, and approach manufacturers on a sure footing.

## REFERENCES

1. *Kempe's Engineers Year Book*, Morgan Grampian, London, 1984, p. F8/2.
2. Hicks, J., *Comprehensive Chemistry*, Macmillan, London, 1981, p. 337.
3. Lambert, L. G., *Instrument Practice*, **19** (1965) 128–37.
4. de Saussure, H B, *Essais sur l'Hygrometrie*, Neuchatel, 1783.

# CHAPTER 2

# *On-line Moisture Measurement in Gases*

While air is the most common gas in which moisture is determined, there are a large number of applications involving other gases, ranging from relatively inert gases such as argon and hydrocarbons to corrosive ones such as chlorine, hydrogen chloride and ammonia. The target concentrations too can vary from a part per million or less in feedstock for ethylene crackers to the near saturated conditions at the outlets of baking ovens. To meet these varied requirements, a correspondingly wide selection of sensors has become available, with often considerable differences in design and performance within a single type. Thus, choice is not straightforward, either of type or manufacturer, since application experience rather than the ultimate in suitability may be the most acceptable criterion for selection.

Water vapour is present in the atmosphere at concentrations of about 10 000 ppm. Hence, meticulous care must be taken in most measuring systems to avoid water vapour leaking or (less obviously) diffusing into the sensed volume and causing grossly wrong results. In addition, water is very strongly absorbed onto most surfaces and therefore such factors as contamination, 'memory' and loss must be at the forefront in the design of any on-line gas handling system for water determination. These factors will be examined later in greater detail.

## 2.1 UNITS OF MEASUREMENT

A number of different ways of expressing moisture content in gases are used. It is necessary to be clear about what is actually being employed and its limitations and particular value.

The commonest ones are:

1. Volume Concentration. This is expressed in per cent or parts per million (ppm) and is the ratio of the volume occupied by the water vapour to the total volume. It is widely used in the petroleum and chemical industries. If ideal gas behaviour can be assumed, it does not depend on the pressure or temperature of observation unless, of course, condensation occurs. Water vapour is not an ideal gas, but, since it is often present in small concentrations in gases such as air which are reasonable approximations to an ideal gas, the errors involved in such assumptions are not large. Nevertheless, the point should be watched if high precision is required or when high moisture concentrations, or polar gases, are encountered.
2. Weight Volume Concentration. This can be given in $g\,m^{-3}$, $g\,litre^{-1}$, $grains\,ft^{-3}$ or pounds per million standard cubic feet. The latter concentrations have been much used in the iron, steel and gas industries. The value depends on the pressure and temperature of the measured gas.
3. Relative Humidity. This is the ratio, expressed as a percentage, of the concentration of water vapour in the gas to the saturated concentration at the same temperature and pressure. Being a ratio it is independent of the pressure and temperature of the measured gas—but does require an accurate knowledge of the saturated water vapour concentration and its temperature and pressure. Water activity of a material is the ratio of its equilibrium water vapour pressure to that of pure water. Thus, a material of water activity 0.5 is in equilibrium with a gas having a relative humidity of 50 per cent—see Section 5.2.9.
4. Dew Point and Ice Point. These are the temperatures at which moisture or ice just condense from a gas as it is cooled under 1 atm pressure. Below 0°C ice is usually deposited, although supercooling can occur. The value will vary as the pressure of the sample gas varies, increasing pressure causing increasing dew point.
5. Wet and Dry Bulb Temperatures. This method measures the temperature depression between ambient temperature and that of a thermometer surrounded by a pure water saturated 'wick' of cloth or ceramic when air is passed over it at a certain range of speeds. This depression is a function of relative humidity.

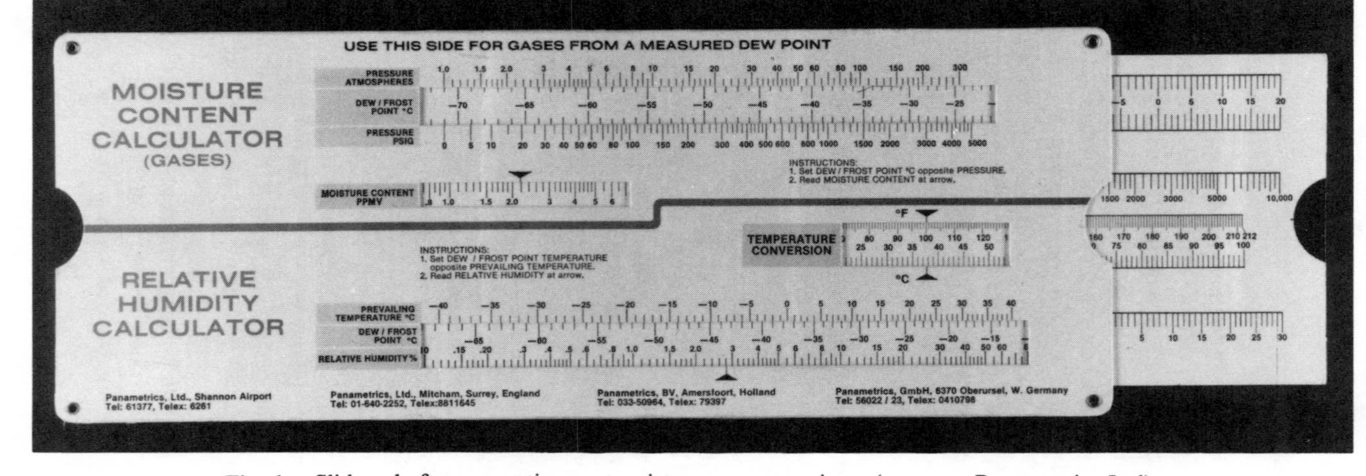


**Fig. 1.** Slide rule for converting gas moisture concentrations (courtesy Panametrics Ltd).

It is worth noting that some sensor manufacturers—for example Panametrics and Shaw—supply slide rules or nomegraphs which allow ready conversion from one value to another (see Fig. 1). International secondary standards for moisture content are generally expressed in terms of dew point and highly precise instruments have been developed for this purpose.

6. Molar Fraction. This is the number of moles (gram molecules) of water present in the mixture ratioed to the total number of moles of all constituents of the mixture. It is used mainly in scientific work. It is obviously independent of pressure and temperature.

## 2.2 SAMPLING SYSTEMS

It is preferable that any gaseous moisture sensor is mounted directly in the pipe, tank or reactor where the determination is required, since this eliminates errors introduced by a sampling system and gives rapid response. The active component of the sensor should be well removed (at least 5 cm) from the containing wall and placed in a part free from any risk of drained condensed water, exits to other unit processes, or minor leaks to the atmosphere. In a pipe, if the diameter is too small to accept the sensor head, then an enlarged section may be used, made of materials with low surface water hold-up. In dust-contaminated streams it may be possible to provide a simple shield such as a sintered metal or plastic hood for the sensor head, to prevent dust build-up or sensor erosion. However, in many applications the sensor cannot be mounted directly in the process stream, due to reasons such as excessive temperatures, high flow rates, pressure, entrained solids, liquids or potential hazard. In these, a sampling system has to be built to carry a representative and uncorrupted sample to the sensor, and return or dispose of the sampled material to a safe place.

The sampling point itself should be well away (typically at least 5 cm) from the pipe or vessel wall, since gaseous composition adjacent to the wall may be non-representative due to adsorption or stagnation. A wide variety of probes is commonly available[1], with facilities to back blow in the event of blockage, construction with corrosion-resistant alloys and the ability to withdraw for cleaning without interrupting plant operation. However, many of these have not been designed with water determination in mind, and their suitability, especially in terms of materials of construction, must be carefully checked against the criteria discussed in the next few paragraphs.

In designing sampling systems for moisture determination, there are a number of basic rules that must be conscientiously obeyed. First, any sampling line should be kept as short and as simple as is possible—extra length delays the sensor response and increases the possibility of adsorption or desorption of moisture. Simplicity reduces the number of components and joints— which reduces the possibility of inward leakage of moisture and contamination. The second basic rule is to ensure that the system is not subject to wide fluctuations in temperature—for these can cause variations in the amount of water adsorbed on the system's inner walls. The third is to keep the whole system at a temperature well above (typically 10–15°C) the highest expected temperature at which water can condense from the gas. Electrical tracing is the obviously preferred method, although steam tracing has been used, preferably with separate coils, not a co-axial jacket with only a single wall between the steam and the sample stream.

Materials of construction of all the sampling system components in contact with the sample stream are vitally important. Plastics and elastomers are not suitable for they take up and emit water over long periods. PTFE is an exception, but even this can show significant permeability to water vapour when very low concentrations are being determined. Care must be taken to overcome its propensity to creep if it is used as a jointing material; as will be constantly stressed, minor leaks to and from the atmosphere are a major source of error. Metals such as brass, copper or aluminium can be used for higher concentrations (dew point $> -25$°C) provided they are adequately cleaned. For levels $< -50$°C dew point carefully cleaned and dried stainless steel or nickel are essential. This would normally involve flushing with a water and residue-free organic solvent to remove grease traces, and then blowing dry for up to 5–10 h with air which has been predried by passing through a molecular sieve column. The same operation may have to be carried out periodically if the process stream itself carries significant grease or oil contamination into the sampling system.

The sample stream may need conditioning before the moisture measurement can be made by filtration or by having its temperature or pressure reduced. Filtration is best carried out with a sintered stainless steel or nickel filter, with a final filter rejecting all particles above 10 $\mu$m. Small cyclones may be used for particle or droplet removal if the flow rate is sufficient. Pressure reducing valves must be suitably constructed with metal diaphragms. If the pressure of the gas is reduced, it will cool, and this may necessitate warming of the gas to compensate for this effect.

After pressure reduction, the gas pressure may have to be measured if it is not at the atmospheric value. Here again a sensor must be chosen which only allows suitable materials to be in contact with the sample system. Where, as part of the sample conditioning, the gas has to be cooled, it is essential to ensure that it is not overcooled so that it approaches or becomes saturated with water vapour. As already pointed out, it is generally recommended that a temperature difference of at least 10–15°C is maintained between the temperature at which moisture would condense from the gas (dew point) and its temperature at the measuring point. Cooling may be by pressure reduction or directly with air or liquid heat exchanger—the simplest being air finning. Liquid cooling coils or jackets, which are not so subject to daily and seasonal temperature variations as air, are preferable, since it is easier to maintain approximately constant temperatures in the temperature reduced sample line to minimise the adsorption and desorption problems already described.

It is obviously essential to ensure that there are no leaks in the sample line, since the omnipresent atmospheric water vapour will cause major errors, especially when determining lower moisture concentrations. It must not be assumed that because the sample line pressure is well above ambient, atmospheric or other surrounding water vapour will not cause problems. Back diffusion of gases into pressurised systems against the outward gas flow is a quantifiable phenomenon. A simplified equation[2] relating to relative concentrations across a leak of length $x$ cm with a streaming velocity $V$ is

$$\frac{C}{C_0} = e^{-Vx/D}$$

where $D$ is the diffusion coefficient. For water vapour in air this is approximately 0.22 at normal temperatures. For example, to maintain a thousandfold difference in concentration across a 0.2 cm leak path, a streaming velocity of 6.9 cm $s^{-1}$ is required. Assuming a leak area of 0.0001 $cm^{-2}$, this is equivalent to a leak rate of 0.007 $cm^3\,s^{-1}$. While this treatment is a gross simplification, it can be used in designing systems as an indication of the maximum leak rate permissible, given existing moisture concentrations and pressure differences. Since this back diffusion rate into a sampling system is roughly constant, a clue to its presence is the variation of the measured moisture concentration with sample flow rate. If the sample stream is discharged into the atmosphere or other moisture-containing dump, it is also important to guard against back diffusion up the sample line. As a guide, at least 2 m of sample line should be present between the sensor and the atmospheric or other discharge point, and a simple non-return valve at the outlet would reduce the chance of moisture diffusing further if abnormal conditions of reduced pressure could occur in the sampling system.

It is good practice to use the pressure of the sampled line or vessel to maintain the flow of gas through the sampling system. Where the sample stream has to be returned to the process, a suitable pair of points, having adequate pressure differential either by use of an existing constriction or deliberate insertion of an orifice plate or similar device, should be selected to give the self-propelled flow. However, in some cases, pumping of the stream cannot be avoided, and this presents some problems. It is very difficult to find a pump in which only clean oil or grease-free metal surfaces or PTFE contact the sampled gas. Hence the pump normally has to be placed well after the moisture sensor. As the use of a pump implies that the sampled line or vessel is, in most cases, at or near atmospheric pressure, then the sampled line will operate at pressures below atmospheric. This accentuates the problems with leaks from the atmosphere. In many applications high concentrations of moisture and water may be present in process lines during abnormal conditions, and

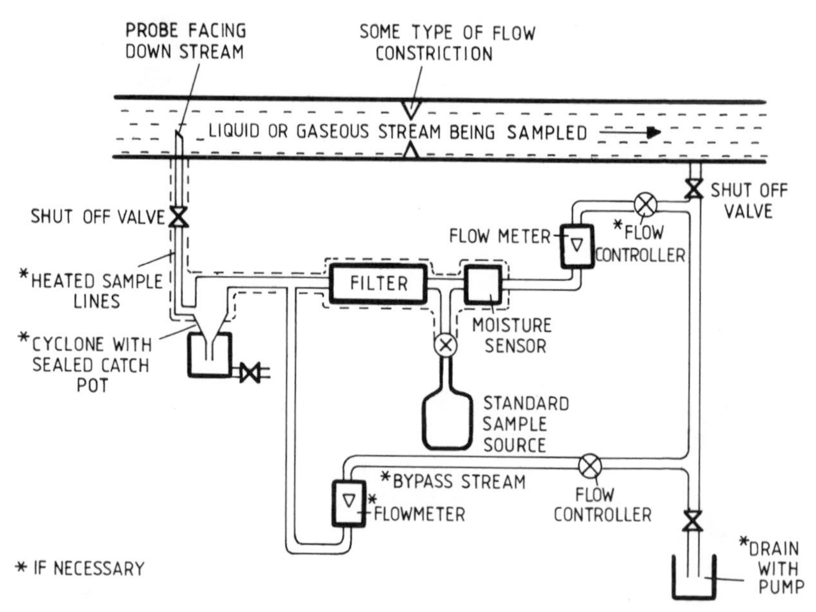


**Fig. 2.** Typical gas or liquid sampling line.

in these cases adequate means of ensuring these do not reach the moisture sensing system— such as drainable catch pots—should be incorporated.

A diagram of a typical sampling system for a lightly dust-laden gas stream is given in Fig. 2. The problems associated with detailed design of sampling systems can be complex, often involving standardisation as well as sample integrity. The reader is referred to standard works on the subject[1] which go into greater detail about such factors as the range of sampling line valves available, alternative means for the removal of fine aerosols and the choice of materials of construction for use with corrosive gases.

## 2.3 SENSOR TYPES

These will be described under the method of measurement of the moisture.

### 2.3.1 Mechanical Sensors

Mechanical moisture sensors depend on the change in length of a sensitive material such as human hair, silk or a specially developed polymer. Rolled human hair with an elliptical cross section gives an enhanced performance, but is mechanically less strong. The change is amplified mechanically and displayed by a pointer against a scale normally calibrated in relative humidity. With some sensors the movement is detected electrically or it can actuate a microswitch and the output is displayed remotely. The output is non-linear with relative humidity, there is considerable hysteresis, slow response (typically 3 min or longer), very poor working below about $-10\,^{\circ}\mathrm{C}$, and the change in material property with exposure conditions—especially in very dry or very wet atmospheres—can result in shifts in the calibration graph. However, these sensors are inexpensive, simple, unpowered and are suitable for non-exacting applications such as factory humidity control in textile manufacture.

Their main advantages are that they:

(a) need no electrical power (in their mechanical form)
(b) are simple and inexpensive
(c) are relatively insensitive to temperature within their working range.

Their disadvantages are that they:

(a) are subject to drift and hysteresis
(b) give a non-linear output
(c) operate over only a limited temperature range.

### 2.3.2 Wet and Dry Bulb Sensors (Psychrometers)

In these sensors two temperatures are measured—those of the sample gas and of a porous substance saturated with pure water and in equilibrium with the sample gas which is flowing over it[3]. The depression of temperature of the porous material can be correlated with relative humidity, the equation relating the values being:

$$P = P_w - 66 \times 10^{-5} B(t - t_w)(1 + 115 \times 10^{-5} t_w)$$

where $P$ is the vapour pressure of water in the sampled gas, $P_w$ is the saturated vapour pressure of water at temperature $t_w$, $B$ is the atmospheric pressure, and $t$, $t_w$ are the dry and wet bulb temperatures, respectively, in °C.

This equation is only valid if the thermometers are in a gas stream of velocity greater than $3\,\text{m sec}^{-1}$. At lower velocities, less well-defined relationships apply. The sensors are widely used in meteorology, with the thermometers placed in a hand-held vane which is rapidly rotated to achieve the necessary velocity. In industrialised versions, platinum resistance thermometers are used, with the 'wick' casing being of porous ceramic fed from a reservoir of distilled water—the latter reduces the frequency of routine maintenance. Calculation of actual relative humidity can be made using a microprocessor, but the simple depression of temperature can often be used effectively for drier control.

The advantages of this method of measurement are:

(a) measurement of relative humidity is made directly
(b) it is simple, inexpensive and its theoretical basis well established
(c) it provides gas temperature as well, which can be of value in drier control.

The disadvantages are:

(a) the rate of evaporation from the wick can be reduced by deposition from the gas stream
(b) the measurement is flow rate dependent below about $3\,\text{m s}^{-1}$
(c) the reservoir requires regular filling with distilled water.

Their main industrial application has been in the control of drying processes and other high humidity applications.

### 2.3.3 Dew Point Sensors

The amount of water vapour that any non-reacting gas—including air can hold depends on its temperature. The higher the temperature, the greater the amount of water vapour that can be present, since the saturated vapour pressure increases with temperature. The dew point is the temperature at which the selected sample of gas is saturated, and any lowering of the gas temperature below this causes the excess moisture to be deposited as dew on any surface in contact with that gas, which is in thermal equilibrium with the gas. Dew point sensors measure the precise temperature at which this dew (or frost if the gas required cooling below 0°C) is deposited on a surface at a closely controlled lowered temperature[4]. The dew point is independent of the temperature of the gas, but it does depend on its pressure. It is used in a special form as a secondary standard for accurate humidity measurements.

The sensor head normally consists of a mirror mounted on a thermoelectric cooler and covered by a cell in which are positioned a light source and detector (Fig. 3). The light from a photodiode or underrun filament bulb is directed at the mirror and the reflected light intensity measured by a photodetector. The light measured may be specularly or diffusely reflected, depending on the make. One model senses the change in capacitance of a cooled ceramic surface when the water condenses upon it (Fig. 4). In optical models the mirror can be made with a gold, rhodium, platinum or specially electroplated nickel surface. It is closely coupled to a two or three stage thermoelectric cooler, which can give temperatures as low as −50°C with air cooling and −75°C with water cooling of the heat sink of the thermoelectric cooler. A control accuracy of better

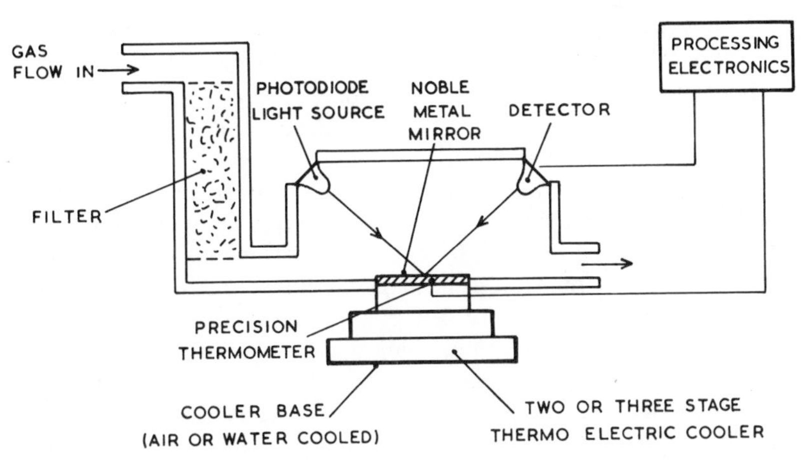


**Fig. 3.** Dew point sensor with mirror detection.

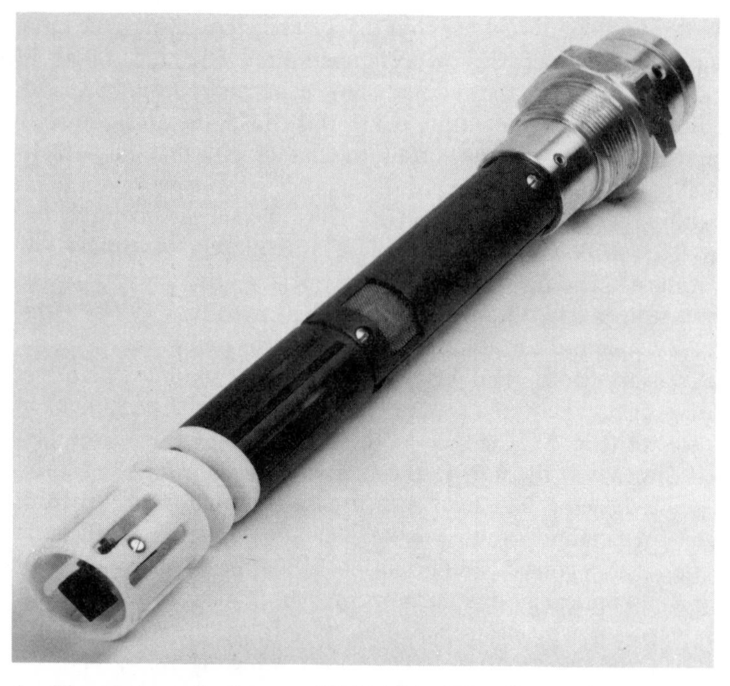

**Fig. 4.** The Endress & Hauser DT 13 Thin Film Sensor for use with the Hydrolog WMT 170 measuring high humidity with dew points up to 170°C.

than ±0.1 °C can be achieved, the temperature being sensed by a platinum resistance thermometer placed in the mirror block. The coolers used are capable at best of giving a temperature differential across them of about 90 °C; hence the lower the temperature of the heat sink, the lower the cooling limit of the mirror or capacitance element. The gas to be measured flows over the mirror at rates between 0.25–2.5 litre $min^{-1}$ through a fine (<25 μm) sintered metal filter, or can be allowed to diffuse to it through a similar filter. In dusty conditions, a prefilter and possibly an after filter on the exit line (if it is short) may be necessary. Various cooling strategies are used, depending on the make of sensor. Some maintain the mirror just at the dew point by feedback control from the optical or capacitive sensor, and continuously monitor the temperature at which dew starts to form, while others cycle through a wider temperature range and detect the temperature of the step produced in the reflected intensity against temperature relationship at the dew point. Advantages are claimed for both techniques; with the former technique the dew point is followed continuously, while the latter can raise the mirror periodically to ambient, which removes other volatiles condensed on the mirror, and requires a lower power dissipation. Variations in the basic design include the use of a second, reference, cell with independent optical systems to compensate for mirror reflectivity changes with the continuously monitored type, and an intermittent cycling method to try and ensure any soluble salts reaching the mirror are re-deposited as larger crystals, not as a finely divided layer over the surface. The effect of dust deposited on the mirror from the gas stream is to reduce specularly reflected light and increase the diffusely reflected light intensities measured by the detectors. This can be compensated for by the two mirror method cited above—although this makes the assumption that deposition on the two mirrors is essentially equivalent. Going through the cooling/heating cycle each time the measurement is made (at the step change in intensity) automatically compensates—provided sufficient light reaches the detector to give a noise-free signal. Continuous sensors use a periodic heating cycle where the mirror is taken to a temperature above ambient and the gain on the photodetector amplifier automatically adjusted to compensate for change in intensity. Dust has only a very small effect on the capacitance type sensor, accounting for its use in dust-contaminated gas streams.

A more subtle effect is caused by the deposition of soluble salts—such as sodium chloride—or other hygroscopic materials upon the mirror. These dissolve in the dew and may raise the temperature at which it evaporates or, if hygroscopic, induce dew formation at temperatures above the dew point. Beyond ensuring adequate filtration of the incoming gas, that no backflow or particulates up the stream is possible, and regularly cleaning the filters and the mirror—washing the latter in microfiltered demineralised water is generally recommended—there is little else that can be done to compensate for these effects. It is claimed that the technique described above which produces big crystals rather than finely divided ones is of value in reducing such interferences but no independent comparative data from a process control environment are known.

While the mirror temperature can be measured to an accuracy of 0.1 °C or better, the accuracy achieved in dew point determination is less than this. Manufacturers claim between $\pm 0.15$ and $\pm 0.2$ °C accuracy in determining the dew point, but this is highly dependent on the state of the mirror. In on-line industrial use a figure of $\pm 0.5$ °C is probably not excessively pessimistic. At a dew point of $-10$°C, and normal atmospheric pressure, this corresponds to an approximate error of $\pm 100$ ppm at the 2600 ppm by volume level—that is $\pm 4$ per cent. Precision may be better than this, but as the condition of the mirror will change with time, drift may be present. The drift in indicated concentration should not be confused with the shift in detected light intensity due to fine dust which is readily compensated for automatically by a calibration check cycle and gain adjustment. The effect is caused by the change in temperature of moisture globule deposition due to the presence of dissolved salts or other hydrophilic centres on the mirror surface as described above.

Most manufacturers supply flow cell attachments for on-line use and outputs suitable for control purposes. Because of the high current required for the thermoelectric cooler (1–5 A) it would be difficult to construct a sensor in intrinsically safe form. Explosion-proofing is possible, but care has to be taken to ensure ready access for mirror cleaning. Response times for the controlled temperature types is typically $2\,°C\,s^{-1}$, above 0 °C, becoming longer at lower temperatures—several minutes for example at −70 °C. Cells are available that will operate at pressures up to 2.1 MPa. Condensable or corrosive gases can interfere—these include $NH_3$, $SO_2$, $SO_3$, HCl, $Cl_2$ and many hydrocarbons, although some corrosive gases can be sensed if the dew point is low enough.

The advantages of dew point sensors are that:

(a) they are a secondary standard and need no calibration except for the most precise work
(b) they are precise and capable of good accuracy
(c) their operating range is very wide—from saturation down to 1 ppm by volume.

Their disadvantages are:

(a) the mirror, when used, requires regular and reproducible cleaning
(b) they are not suitable for use with condensable gases
(c) they require a highly filtered gas stream to minimise salt errors and give long periods between minor maintenance.

They have been used widely for the determination of moisture in foodstuffs by the equilibrium relative humidity method, for sensing the onset of dew in storage systems, and for dryer and cooking oven control. In the latter cases the capacitance type has proved especially valuable due to its insensitivity to dust and grease contamination. Personally communicated industrial experience indicates that both the mirror and capacitance sensors can show a high level of reliability and accuracy when used on line, with good accuracy at all moisture levels. However, there appears to be considerable difference in performance from one manufacturer to another with the mirror-type sensors, especially at higher moisture concentrations.

### 2.3.4 Electrolytic Sensors

These employ phosphorus pentoxide placed between two noble metal electrodes which have a selected DC voltage difference maintained between them[5]. The pentoxide is an extremely hygroscopic material, which forms phosphoric acid on absorbing water. When water is absorbed the hydrated pentoxide becomes conducting and electrolysis takes place due to the voltage difference between the electrodes. Oxygen and hydrogen are evolved from the electrodes and in theory the pentoxide is restored. The resultant electric current between the electrodes depends on the rate of absorption of water vapour from the gas stream. Because the water is dissociated on measurement the sensor must operate in a gas stream flowing at a constant rate. Typically this will be $100\,cm^3\,min^{-1}$, controlled to $\pm 2$ per cent. Commercial sensors can cope with pressure up to 690 KPa, and sample gas temperature can vary between 0 and 80°C, but performance is improved if it is controlled to about $\pm 1$°C. Quoted accuracies range between $\pm 5$ and $\pm 2$ per cent, with a 90 per cent response to step change in 1 minute. Ranges vary between 0–10 and 0–3000 ppm by volume, with limits of detection of less than 1 ppm. The measuring cell needs regenerating or replacing from time to time—lives in excess of $250\,000\,ppm\,h^{-1}$ (roughly 100 days at 100 ppm) are quoted. Since it is based on a well-defined chemical reaction, calibration is not normally required. However, there is some question as to whether the reaction goes to completeness in the reverse direction, and whether the degree of completeness is a function of the age of the sensor. Problems can also arise with the catalytic recombination of the hydrogen and oxygen formed, but this effect can be markedly reduced by the cell design and the choice of electrode material—rhodium giving an improved performance relative to platinum. Gases to avoid are corrosive ones such as 'wet' chlorine and hydrogen chloride, those like alcohols—that combine with the pentoxide—and obviously alkaline gases such as ammonia and amines which neutralise the acidic pentoxide. Unsaturated hydrocarbons polymerise on the oxide forming a solid or liquid coating which clogs the cell. The probe is commercially available in explosion-proof form for hazardous areas.

Electrolytic sensors have the following advantages:

(a) they have low limits of detection
(b) they do not require calibration.

Their disadvantages are:

(a) they require a constant gas flow rate
(b) the pentoxide is highly reactive, which limits the range of gases with which it can be used
(c) the cell requires regular regeneration
(d) the cell is destroyed by accidental water immersion.

Applications quoted include the determination of moisture in natural gas, 'dry' chlorine, transistor filling gases, feed gas in polyethylene manufacture, refrigerant gases, heat treatment gases and sundry applications involving determining lower levels of water in gases which do not react with phosphorus pentoxide. Reliability and stability problems with these sensors have been personally communicated to the author, and there is a tendency to replace these sensors with the silicon or aluminium oxide type sensors, which are described later (Section 2.3.6).

### 2.3.5 Lithium Chloride Hygrometers

Lithium chloride is a hygroscopic, water soluble salt. When in equilibrium with a gas, its water content will depend on the relative humidity of that gas. A typical sensor using it consists of a thin layer of the chloride with an additive such as polyvinyl alcohol to improve wetting, between two noble metal electrodes connected to an alternating voltage source. The lithium chloride takes up moisture, becomes conducting and allows an alternating current to flow—which does not cause electrolysis but heats up the sensing element to a temperature where the water is driven off and the conductivity drops sharply. This temperature is sensed with a platinum resistance or other thermometer, and is used together with the ambient temperature to calculate the relative humidity of the gas. Absolute humidity or dew point can be deduced from the drying temperature without the need of measuring the ambient temperature. The typical humidity range covered is between 10 and 100 per cent at 20°C, with a dew point range of $-40$ to $+90$°C (or up to 130°C in some cases). Measuring accuracy claimed is $\pm 0.5$°C. Response time to an 80 per cent step change is about 2 minutes. While a constant gas flow rate is not necessary, it should be less than 2 $sec^{-1}$ linear velocity. The lithium chloride element needs renewing when its performance deteriorates—this implies regular checks—and if operated consistently above 100°C, it must be re-moistened at regular intervals. Gas pressures up to 1 MPa can be used with commercially available probes. Gases to be avoided include sulphur dioxide, sulphur trioxide vapour, ammonia, high concentrations of carbon dioxide, chlorine, hydrogen sulphide and condensable hydrocarbons.

These sensors have the following advantages:

(a) they are not dependent on flow rate
(b) they can measure dew point and relative humidity.

Their disadvantages are:

(a) the probes need servicing at a frequency dependent on particular application
(b) they are prone to attack from a number of gases
(c) they are destroyed by accidental immersion
(d) they cannot measure very low moisture concentrations.

They have been used for dryer control in the ceramic and textile industries and in hot and cool plant air in blast furnace operation, but they are steadily being replaced by other probes such as the silicon, aluminium oxide or capacitance dew point meters.

### 2.3.6 Aluminium Oxide Hygrometers

These sensors, which are widely used in industry, consist of a strip, small plate, deposited layer or wire of high purity aluminium. The surface is chemically oxidised to produce a pore-filled insulating layer of partially hydrated aluminium oxide. A water-permeable but conductive gold film is deposited on the oxide layer to form the second electrode of a capacitor (the aluminium metal being the first) with the aluminium oxide as the dielectric (Fig. 5). In situations where aluminium can be chemically attacked silicon sensors may be used as an alternative (Section 2.3.7) with tantalum metal-based sensors being the most inert although their sensitivity is less. The oxide layer is in the form of a mass of tubular

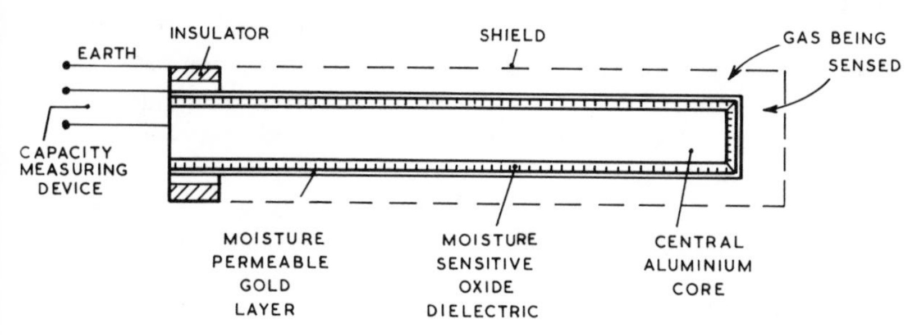


**Fig. 5.** Aluminium oxide sensor.

pores running up from the metal base to the exposed surface. Change in the size of these tubules with time is suspected as being the cause of the slow shifts in calibration often experienced with these sensors. Water is absorbed in these tubules in amounts directly related to the moisture content of the gas in contact with it.[6] The amount of water is sensed electrically by measuring the change in capacitance or admittance produced by this water. Because of the radius of the pores in the aluminium oxide, the sensor is virtually specific for water molecules. The dew point range claimed to be covered by standard sensors is between $-110°C$ and $+20°C$, which corresponds to a range from approximately 0.001 ppm to 0.2 per cent by volume. Some claim to operate up to 100°C but the shifts in calibration tend to be accentuated by higher temperature operation and more frequent calibration checks would be required. Probes can operate over wide gas temperature ranges—from $+100°C$ to $-110°C$. Temperature coefficients are small but need to be taken into account especially in the higher gas temperature and dew point ranges. For accurate measurement in these ranges it is essential that the probes are calibrated at their expected operating temperature. Response times for 63 per cent change is a few seconds at higher dew points, increasing as the dew point drops to a minute or more below $-50°C$ ice point. Commercially available sensors can operate at pressures up to 34 MPa and at vacua down to 0.7 Pa. However it is important to note that the reading obtained is dependent on pressure. A sensor operating at 1 MPa will give the dew point of the gas at 1 MPa, not at atmospheric pressure. If line pressure variations are common, it may be better to take a continuous sample of the gas and reduce its pressure (always ensuring the cooling caused by the gas expansion does not bring its temperature to the region of the dew point) to atmospheric or other controlled value. Accuracies claimed range from $\pm 1°C$ to $\pm 2°C$ at higher dew points to $\pm 2°C$ to $\pm 3°C$ at $-100°C$ ice point. They can tolerate linear flow rates of up to $10\,m\,s^{-1}$ at 0.1 MPa. While their calibrations do drift with time, it is important not to overemphasise the magnitude of these drifts. The stability does vary from manufacturer to manufacturer and many guarantee the original calibration for periods between 6 and 12 months, within the accuracies quoted above.

These sensors are free from interferences from gases such as hydrocarbons, carbon dioxide, carbon monoxide and chloro-fluoro-hydrocarbons, although drift rate may vary in differing gases. Certain corrosive gases, such as ammonia, sulphur trioxide vapour and chlorine, attack the sensing element and should be avoided. Where occasional dust is expected in the line a sintered metal filter should be used to cover the sensor. If heavier concentrations are encountered, it may be necessary to mount the sensor in a sample line with a cyclone or filter to remove the particulates. The performance of the sensor's element is not affected by dust, but access of water vapour to the sensitive surface is hindered. The probes themselves can be intrinsically safe if operated through Zener barriers or galvanic isolators, and can be used in all areas including zone 0.

Aluminium oxide sensors have the following advantages:

(a) wide dynamic range from 0.1 ppm to 100 per cent
(b) they are relatively stable, with low hysteresis and temperature coefficients
(c) they are flow-independent
(d) they can be intrinsically safe
(e) they have high selectivity for moisture
(f) they operate over a wide range of temperature and pressure
(g) they require little or no maintenance.

Their disadvantages are:

(a) the calibrations show slow drifts, which may accelerate at higher operating temperatures or in certain gases
(b) some corrosive gases affect the probe (analogous silicon or tantalum oxide sensors may then be alternatives)
(c) sensors need simple temperature control at high temperature and dew points
(d) the sensors are not absolute and require calibration.

The range of application of these sensors is very extensive. Ethylene feed gas in polythene manufacture, dryer control, natural gas directly in pipelines, lamp filling, hydrogen recycle streams in catalytic reformers and bottled gases just give an indication of the range of materials and uses. They have been well received by a wide range of industries and have been the preferred choice in a very large number of applications particularly at lower dew points. However, the silicon sensor (Section 2.3.7) is expected to make a significant impact on this preference.

### 2.3.7 Silicon Hygrometers

These are capacitive type sensors, in some ways similar to aluminium oxide sensors, but employing silicon rather than aluminium as the sensor material (Fig. 6). They are fabricated using silicon technology and should not be confused with the silicon circuit type detectors that employ an evaporated layer of aluminium, subsequently oxidised to form the sensing element. In their most stable and versatile form the silicon chip is temperature-controlled at above ambient temperature to maintain independence from gas temperatures.

The sensors have an operating dew point range from −80 °C to above 80°C, although the standard calibrated range is −80°C to +10°C. Their stability is such that they can have a dew point discrimination of 0.001 °C. They also show very rapid response—with suitable purging a sensor can shift from being exposed to saturated gas to reliably measuring at 1 ppm moisture level in less than 15 s. They are capable of operating at pressures ranging from vacuum to 25 MPa, and at gas temperatures from −40 °C to above 45°C, without affecting their performance. Because silicon is on the whole a more chemically stable material than aluminium they should have areas of application with some gases where the aluminium sensors are inappropriate.

Silicon sensors have the following advantages:

(a) wide dynamic range from below 1.0 ppm to saturation
(b) high stability
(c) they are flow-independent
(d) they operate over a wide range of temperature and pressure
(e) high selectivity
(f) very rapid response
(g) they are relatively chemically inert—can stand immersion as well.

Their main disadvantages are:

(a) they are relative, not absolute sensors and require calibration
(b) they can show drift when operating in certain gases, e.g. $CO_2$
(c) they are currently more expensive than aluminium oxide sensors.

Their application range is similar to that for aluminium oxide sensors, but even broader due to their superior characterisation. In general they would be preferred to the aluminium oxide type sensor for most applications.

### 2.3.8 Polymer Humidity Sensors

There are a number of moisture sensors using an organic polymeric element which absorbs or desorbs water as the relative humidity of the gas surrounding it changes. The materials used are often proprietary, but include sulphonated polystyrene with the sulphonation confined to the

**Fig. 6.** Si-Grometer moisture sensors (courtesy MCM Ltd).

surface layer, giving rapid response. The impedance or capacity of the sensitive layer is measured and this related to the relative humidity of the surrounding gas. Operating temperatures range between −50 and 125°C with relative humidity ranges between 0.5 and 100 per cent and a discrimination of 0.1 per cent. Accuracy is limited when large humidity excursions occur by the hysteresis, which can amount to ±2.5 per cent relative humidity at mid range. With lesser excursions—say a maximum of between 10 and 20 per cent—an accuracy of ±1 per cent can be achieved. The temperature coefficient can range between 0.5 and 0.05 per cent relative humidity per degree centigrade. Especially with the former sensors the temperature of the sensing element is measured with a platinum resistance thermometer or thermistor and the value used to compensate for this effect. Response time to 90 per cent of final signal is between 1 and 30 s depending on the sensor construction and the size of the moisture concentration step—the latter figure corresponding to a 60 per cent change in relative humidity. Their output is not flow rate dependent. They are more sensitive to chemical interference and attack than the aluminium oxide type sensors—alcohols and amines, for example, can give high readings while aromatic hydrocarbons, acid vapour and acidic oxides such as $SO_2$ and $NO_2$ can be destructive. The interfering gases for a particular sensor depend on the polymer used. Sensors such as these may show drifts in the 80–100 per cent relative humidity area, but these can be minimised by keeping them more or less continuously within this range, running them at a controlled raised temperature and calibrating under these operating conditions.

The advantages of these sensors are:

(a) simple, low cost relative humidity measurement
(b) they are flow rate independent
(c) they require minimal maintenance
(d) they give a rapid response to humidity changes.

Their disadvantages are:

(a) they measure only a limited range of moisture content
(b) they are relatively sensitive to chemical interferences and dust deposition
(c) they show hysteresis and drift at high humidities.

They are widely used to measure relative humidity in instrument rooms, textile mills, tobacco processing, dryers and some other industrial processes. They vary greatly both in quality and price, and with the more expensive a considerable amount of sensor research and applications engineering has resulted in a much more industrially acceptable product.

### 2.3.9 Ceramic Moisture Sensors

These are similar in application and performance to the polymer type sensor, but may have improved performance at extremes of temperature and higher resistance to some inorganic vapours. Essentially they consist of a wafer or deposited thin film of ceramic whose resistivity or capacitance varies with the relative humidity of the gas—normally air—surrounding it. The range of compositions used is quite wide, but formulae based on magnesium chromite with $TiO_2$ and $LiF/Al_2O_3$ have been described.[7]

Their advantages are:

(a) they are relatively temperature-insensitive
(b) the material is more chemically inert than the organic polymers
(c) they show good performance at high humidities.

Their disadvantages are:

(a) they can be sensitive to surface deposition
(b) they show limited response at low moisture concentrations.

They tend to be used as relative humidity sensors for its control in rooms and to a lesser extent dryers. They do not seem to have made major inroads into the market dominated by the $Al_2O_3$ or silicon type sensors.

### 2.3.10 Crystal Oscillator Sensor

This unique sensor determines moisture content on line by measuring the changes in resonance frequency of vibration of a quartz crystal with a hygroscopic coating. The crystal (Fig. 7) is alternately exposed to the 'as received' and dried gas, and the resulting shift in frequency is due to the change in the mass of the crystal plus coating due to water uptake. This is achieved in the following manner. A sample stream is filtered and split into two streams in the measuring head. One of the two streams is dried by molecular sieve, and the streams are periodically switched by valves

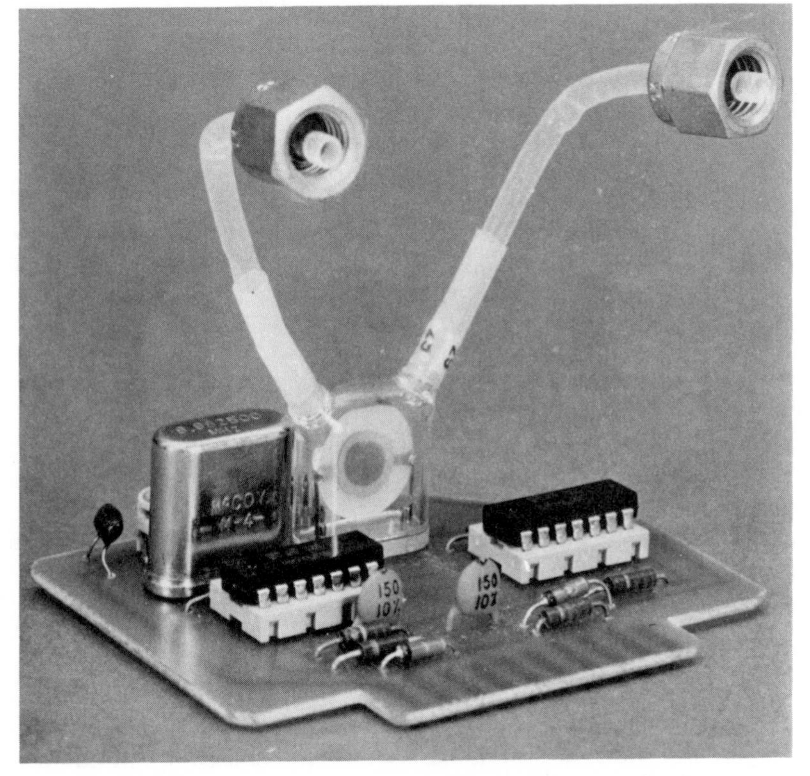


**Fig. 7.** Sensing crystal in oscillator type instrument (courtesy Du Pont Ltd).

so that they alternately pass over the crystal at 30 s intervals, giving a 90 per cent response time in 60 s. The frequencies of oscillation of the crystal in dried and 'as received' gases is recorded, referred to the frequency of a sealed reference crystal and used to calculate the moisture level with a dedicated microprocessor, using a polynomial expression. The crystals are temperature-controlled at 60°C. The sensor's head also contains an internal calibrator that can be used to manually check performance when required, although drift-free operation for periods longer than a year have been reported. The range of the instrument using the microprocessor output unit is from below 0.02 ppm by volume to 100,000 ppm by volume, although the sensor is only calibrated up to 1000 ppm. Accuracy is quoted as 1 ppm or 5 per cent, with 0.1 ppm in the 0.5 ppm range. The pressure of the flowing gas at the sensor is controlled to 0.103 MPa. It is available in explosion-proof form and may be operated in the ambient temperature range of −18°C to +52°C.

Its advantages are:

(a) it has a wide working range
(b) it has a true calibration check 'built in'
(c) auto-zeroing is inherent in the design
(d) it is available in explosion-proof form.

Its disadvantages are:

(a) it is expensive and bulky
(b) it cannot be used in-stream
(c) it has a 1-minute response time.

Applications include those where very high reliability moisture sensing is essential or where very low concentrations ($<1$ ppm by volume) have to be accurately determined such as moisture in ethylene, catalytic feedstocks, natural gas, refrigerants, doping gases and various monomers. Where high reliability is required, an economical alternative may be triplicated silicon or aluminium oxide sensors with the control computer voting on their outputs, although this does not rule out simultaneous deviation due to corrosive and/or interfering gases or long-term parallel drift.

### 2.3.11 Infra-red Moisture Sensors

Infra-red gas analysers can determine gaseous moisture content[8]. In these the transmitted intensity of infra-red radiation at a wave length selectively absorbed by water vapour is measured. This is usually ratioed to a transmitted intensity of radiation at a wavelength or wavelengths not selectively absorbed by water. The wavelength selection may be achieved by interference filters, which use multi-layers of solid dielectric which only transmit over a limited range, by gas filters containing water vapour, or by a luft type detector which is filled with water vapour and senses pressure changes as the infra-red radiation which has passed through the sample stream is absorbed after being 'chopped' by a mechanical shutter. The beam of infra-red radiation may be directed across a duct or pipe, or a sample continuously taken off and passed through a flow cell after suitable pretreatment. The concentrations that can be determined range between a few ppm up to 100 per cent, but this is dependent on what other gases are present and on the wave length used. Response times similarly vary between a fraction of a second up to a minute or more at lowest concentrations. Overall accuracies of $\pm 1$ per cent of range are quoted, and these are achievable in practice with adequate calibration. The limits imposed on the gases that can be examined are due to materials of construction of the cell windows—sapphire being a popular material—or interfering infra-red absorption, for example with some alcohols and ammonia deviatives.

The advantages of the technique are:

(a) it covers a wide range of concentrations
(b) it can be used for multi-component analysis simultaneously
(c) it can be used with highly corrosive gases.

Its disadvantages are:

(a) it can be relatively expensive
(b) a number of gases can interfere
(c) the technique requires calibration.

It is being used to an increasing extent for the determination of moisture, especially in corrosive gases such as hydrogen chloride and sulphur dioxide and where multi-component analysis including water is required.

### 2.3.12 Other Gaseous Moisture Sensors

On-line mass spectrometers[9] and gas chromatographs[10] can determine moisture down to 30 ppm or less. The response time is rapid for mass spectrometers but can be a matter of minutes with gas chromatographs, where adsorption of moisture on the column can cause problems. Water would normally only be determined with those instruments as one part of a multi component liquid. Harrison[11] has described the use of fibre optics in the multi-attenuated total reflectance mode to determine moisture in gases using a 40-m fibre of fluoride glass. No performance figures are quoted, but it does give the possibility of completely power free moisture sensing in remote or hazardous environments—the powered infra-red source and detector being situated remotely from the hazard. An optical fibre sensor using visible light and a cobalt chloride/gelatine sheath has also been described[12]. Many moisture-sensitive integrated circuits have been described in the literature[13] but as yet none have been fabricated in a form that would give the stability and ruggedness required by an industrial sensor. It should be possible to use dielectric constant measurements to determine moisture in air and other relatively low dielectric content gases such as hydrogen or argon, but no application of this has been discovered. Thermal conductivity[14] can also be used to determine the moisture content of gases. Limit of detection of less than 0.01 per cent moisture can be achieved, but the residual gas must remain essentially constant in composition as far as thermal conductivity goes.

## REFERENCES

1. Cornish, D. C., Jepson, G. and Smurthwaite, M. J., *Sampling Systems for Process Analysis*, Butterworths, London, 1981, pp. 25–95.
2. Jost. W., *Diffusion*, Academic Press, New York, 1952, p. 410.
3. Arnold, J. H., *Physics*, **4** (1933) 255-62, 334–40.
4. Wylie, R. G., Davies D. K. and Caw, W. A., in: *Humidity and Moisture*, Vol. 1 (Ed. R. E. Ruskin), Reinhold, New York, 1965, pp. 125–34.
5. Jones, R. H. and Petersen, A., in: *Humidity and Moisture*, Vol 1 (Ed. R. E. Ruskin), Reinhold, New York, 1965, pp. 507–11.
6. Khanna, V. K. and Nahar, R. K., *Sensors and Actuators*, **5** (1984) 187–98.
7. Tofield, B. C. and Williams, D. E., *Solid State Ionics*, **9 & 10** (1983) 1299–302.
8. Willis, H. A., *Advances in Infra Red and Raman Spectroscopy*, Vol. 12, Heydon and Sons, London, 1976, pp. 81–135.
9. Dunn, B. E., *Advances in Instrumentation*, **27** (1972) 736/1–5.
10. Villalobos, R., *Analytical Chemistry*, **47** (1975) 983A–1004A.
11. Harrison, A. P., UK Patent Application 2130739 (1984).
12. Russell, A. P. and Fletcher, K. S., *Analytica Chimica Acta*, **170** (1985) 209–16.
13. Jachowicz, R. S. and Senturia, S. D., *Sensors and Actuators*, **2** (1981/82) 171–86.
14. Cherry, R. H., in: *Humidity and Moisture*, Vol. 1 (Ed. R. E. Ruskin), Reinhold, New York, 1965, pp. 539–51.

# CHAPTER 3

# *On-line Moisture Measurement in Liquids*

Water in liquids exists in the following forms: (i) dissolved in widely varying concentrations in miscible or near miscible liquids such as acetone ethanol or glycol, (ii) dissolved at very low concentrations in immiscible liquids such as paraffin, transformer oil or refrigerant, or (iii) as droplets in varying degrees of dispersion in water saturated immiscible liquids. The first two cases are generally amenable to standard sampling and sensing techniques while the third case may present problems. For the determination of water in single-phase liquids, infra-red, aluminium oxide, refractive index, capacitive or conductivity sensors can be used. Multiphase liquids can also use light scattering, while certain physical techniques such as density are applicable to both. Concentrations in liquids are normally expressed on a simple weight/weight basis, ranging from ppm to per cent. Volume/volume or weight/volume are very rarely used.

## 3.1 SAMPLING SYSTEMS

Where the sample is a homogeneous liquid the sampling system will consist of a probe in the sample stream (sited well away from the wall of the vessel or pipe to avoid anomalous concentrations that may exist by the wall), valves to control the sample flow and isolate the sensor for maintenance and in some cases calibration, and components to filter or degas the sample stream. The flow through the sample system may be achieved by pumping, gravity or inducing a pressure differential in the main flow to cause flow through the sampling system. It is important to remember the ease of adsorption of water due to its high polarity, stressed in the previous chapter, when determining lower concentrations of water in liquids, and bear this in mind when choosing materials of construction for the sampling system. In general the same rules apply for liquid and gaseous systems.

When determining the water content of multiphase liquids, they first require representative sampling, although it is sometimes possible to avoid this by obtaining reproducible dispersion of the water droplets with jet mixing or ultrasonic/cavitational techniques and then measure light absorption or scattering on the whole stream before and after dispersion, or measure scattering at differing angles to eliminate the effect of suspended solids[1]. Where droplets are present, whether the water sensor operates in stream or on a sample stream taken from it, it is essential that the sensor sees a representative portion of the whole stream. This itself may involve droplet dispersion, high turbulence or isokinetic sampling, followed perhaps by degassing or filtration—care being taken that neither operation removes significant amounts of water. The task of representative sampling of heterogeneous sample streams is a complex one and the reader is referred to standard works on the sampling of liquids for analysis[2]. A typical sampling system for a liquid containing entrained gases and solids is shown in Fig. 2.

It is also possible to determine the water content of a homogeneous (but not a heterogeneous) liquid from the relative humidity of the gas above the liquid, when it has reached equilibrium at a known temperature—the so-called equilibrium relative humidity (ERH) method. The relative humidity of the gas can be determined by a sensor selected from those described in the previous chapter. Since many liquids can be vaporised readily, it is possible to carry this out first on an isolated representative sample or small sample stream and again use a gaseous moisture sensor to determine the water content. Corrosion rates are often highly dependent on trace moisture and in some difficult applications, sensing of corrosion rates galvanically may be a viable way of inferring the presence of traces of moisture.

## 3.2 SENSOR TYPES

These will be described under the method used to determine the water content.

### 3.2.1 Infra-red Sensors

These measure the absorption of infra-red radiation at wavelengths that are strongly and selectively absorbed by water—its so called 'characteristic wavelengths'. These are normally 1.45, 1.94 or 2.95 $\mu$m[3] although longer wavelengths can be used in especial circumstances such as wavelength interferences. The basic components of the sensor head are infra-red source, chopper and/or interference filter holder, flow cell, focussing optics and a detector. The source may be an electrically heated nichrome wire, ceramic rod or quartz-iodine light bulb, depending in part on the wavelength used. The chopper serves to alternately rotate two or more narrow wavelength band interference filters across the infra-red beam—one filter transmits radiation at a wavelength that is strongly absorbed by water, while the one or more remaining ones transmit at wavelengths which are adjacent to this but not absorbed by water. Chopping frequencies are relatively rapid for a mechanical system—typically 70 Hz. After passing through the cell containing the flowing sample stream the intensities of the radiation at different wavelengths are measured by a photo-conductive or pyro-electric detector. A remotely sited read-out unit separates the intensities transmitted by the two or more filters, then calculates the absorbance and hence the water concentration in the material analysed. Correction for the effect of an interfering component can be made at this stage, provided a filter sensitive to that component has been included in the filter wheel.

The sample continuously flows through a cell with infra-red transmitting windows. One of the general problems encountered with the infra-red analysis of liquids is that very short cell lengths (typically 30–300 $\mu$m) are required for longer wavelength operation or, with highly absorbant liquids such as water, to allow sufficient radiation to be transmitted for precise measurement to be attained. In flowing through such narrow gaps the process fluid and its contents must not lead to blocking, or window fouling, for it is practically impossible to clean these windows without taking the cell apart. With shorter infra-red wavelengths (approximately 0.8–2 $\mu$m) or infra-red transparent liquids, such as carbon tetrachloride, longer cells (0.1–10 cm) can be used since the absorbance is less. The use of the shorter wavelengths can however often result in poorer limits of detection albeit a better sensitivity at higher concentrations. A viable alternative is to use a technique called multi-attenuated total reflectance (MATR or simply ATR). In this variation of the method the infra-red beam passes by multiple internal reflections along an infra-red transmitting crystal, such as silicon or zinc selenide, which is immersed directly in the process stream or in a relatively large flow rate 'fast loop'. Absorption at the characteristic wavelengths of all components in the liquid occurs at every internal total reflection—as if the beam penetrated a few microns into the process fluid at each reflection. This eliminates the need for narrow cells and permits easy cleaning, large flow rates through the analyser and the direct determination of water in slurries or viscous liquids.

Limits of detection of water with cell or ATR type liquid infra-red analysers range from 1–100 ppm depending on the application. The upper limit of measurement approaches 100 per cent, although accuracy, which at best approaches $\pm 1$ per cent, is poorer at higher concentrations. A wide range of liquids can be examined, the main limitation being compatibility with the infra-red transmitting window or ATR crystal materials. Some such as calcium fluoride and quartz are relatively inert, but are useable over a limited wavelength range, while others such as zinc selenide and silver chloride cover a wider wavelength range but are more reactive. However there are few problems where lack of a suitable material is the limiting factor in the application.

The advantages of this analytical method are:

(a) it is selective
(b) it covers a wide range of concentrations
(c) it has fast response—typically 0.5 s to 90 per cent response
(d) it can be made highly corrosion resistant
(e) it is available in explosion-proof form.

The disadvantages are:

(a) the sample stream may need filtration or other conditioning
(b) the performance is very application dependent
(c) it cannot readily be made intrinsically safe
(d) it is relatively expensive.

The technique has been used to determine water in fuel oils, petroleum derivatives, acetic, nitric and sulphuric acids, alcohols, deuterium oxide, liquid ammonia, chlorinated hydrocarbons, vegetable oils and many chemical intermediates. ATR is normally used with liquids containing major amounts of water and acids, and has also been successfully applied to determining non-dissolved, dispersed water in fuel oils. This gives a two-part calibration graph—the slopes changing above the saturation level when suspended water droplets are present.

### 3.2.2 Aluminium Oxide Sensors

These sensors have already been described in detail in the previous chapter on gaseous moisture sensors (Section 2.3.6), where they have their main areas of application. Some manufacturers' probes may be directly immersed in liquids to determine their dissolved water content. It should be noted that the analogous silicon sensors cannot currently be used in this way (Section 2.3.7). In a liquid water molecules diffuse through the gold layer on the outside of the probe and reach equilibrium in the pores of the aluminium oxide layer. The quantity of water molecules absorbed alters the dielectric constant of the aluminium oxide, which is continuously monitored. Since the pores in the oxide layer are smaller than most organic liquid molecules, the molecules cannot penetrate them. The sensor acts as a semi-permeable layer, allowing the measurement of water vapour pressure in organic and some other liquids. Henry's law is used to relate the measured vapour pressure—which in turn determines the water uptake in the aluminium oxide—to the concentration of dissolved water in the liquid. The law states that the mass of a slightly soluble vapour that dissolves in a given mass of liquid, at a given temperature, is very nearly directly proportional to the partial pressure of that vapour. Thus when the saturated vapour pressure for water in an organic fluid is known, the constant of proportionality can be calculated and water concentrations determined for other vapour pressures. If the saturated water concentration in the fluid is not known, it is possible to estimate it by relating the water vapour pressure measured by the probe to that of water at the same temperature. Where greater accuracy is required or measurement is necessary at higher water concentrations with fluids in which water is more soluble, an empirical calibration should be made using well-dried fluid (a molecular sieve or the highly reactive phosphorus pentoxide take water concentration in most cases down to less than 1 ppm) and known additions of dissolved water. Care must be taken to ensure that the materials contacting these standards do not absorb water from them, that the added water is fully dissolved and that the standards are, at low concentrations, prepared and stored under dried nitrogen or other water free gas in water-impermeable and non-adsorbing containers.

These sensors will determine from saturation levels of water (a maximum of about 1 per cent) down to $5 \times 10^{-12}$ in favourable cases such as aliphatic hydrocarbons. For water in straight chain alkanes the constant of proportionality is essentially independent of temperature. However, for many other liquids the constant is a function of temperature. For water in aromatics and olefines the constant increases by about 3.5 per cent per °C over a range between 0 and 50°C —and correction must be made for this. The sensors can operate over a range of about +70 to −100°C and at pressures up to 34 MPa (5000 psig). Flow rate limits of less than 5 cm $s^{-1}$ are suggested in liquids—to avoid mechanical damage—with response times of the order of 30 s in adequately moving systems. The probe is covered in a sintered bronze or PTFE filter to avoid minor solids deposition, but if higher solids are present than the sensors must be placed in a sample line with adequate filtration facilities.

Accuracy obtainable is dependent on sensor stability, calibration and/or the applicability of Henry's law. The dew point accuracy achieved by these sensors is $\pm 2$°C—which corresponds to between 128 and 164 ppm water with a Henry's law constant of approximately 6 and a 'dew point' of 25°C. Reproducibility is $\pm 1$°C with a corresponding increase of precision—between 148 and 154 ppm in the cited case.

The advantages of these probes when used with liquids are:

(a) they are highly selective for water molecules in a wide range of organic liquids
(b) they have a wide dynamic range (11 orders of magnitude!)
(c) they have direct in-line measurement at high pressures
(d) they can be made instrinsically safe
(e) measurement is rapid.

Their main limitations are:

(a) they only work successfully at lower concentrations
(b) the probe can readily be corroded by a number of liquids or dissolved gases—$H_2S$, HCl for example.
(c) an upper limit of operation of 70 °C
(d) limited accuracy.

On-line applications include the determination of water in coolants, diesel fuels, aromatics, liquid $CO_2$, transformer oils and a range of organic intermediates.

### 3.2.3 Equilibrium Relative Humidity

This technique is widely used in determining the moisture content of solids, and the immersed aluminium oxide sensor described above is indeed a special case of this. It assumes that a certain volume of gas—say that emerging from a dryer—is in equilibrium with the material in the dryer, and hence its content can be inferred from the moisture content of the gas. It will be discussed more fully in the chapter on moisture sensors for solids (Section 5.2.9) but is equally applicable to relatively non-volatile liquids. One qualification with more volatile liquids is that the moisture sensor used must be insensitive to and not attacked by the other liquid components in equilibrium with the gas.

The method's advantages with liquids are:

(a) it is relatively inexpensive
(b) a sampling system is not required
(c) it can be used with aggressive liquids, providing they are non-volatile.

Its disadvantages are:

(a) it requires the gas to be in equilibrium with the sample stream, which due to the smaller surface area of bulk liquids relative to powder, may present problems
(b) it may have limitations when the liquid sample gives off corrosive or interfering vapours.

Few specific applications are known or quoted in the literature, with the exception of the special case of the aluminium oxide sensor quoted above.

### 3.2.4 Physical Property Sensors

Where the liquid can be considered a binary mixture of water and a second component, its water content can be measured by determining a selected physical property of the liquid. Useful properties include refractive index, dielectric constant, density, conductivity and viscosity. They are widely used in industry, especially where water is a major component, as in drinks and mineral acids.

### 3.2.5 Process Refractometers

Process refractometers are usually based on critical angle measurement. They sense precisely the angle at which a light beam just stops penetrating into the liquid and is instead totally internally reflected onto a detector or detector array. A prism of an inert material having a high enough refractive index, such as sapphire, is used as the optical contact, one face being immersed in the process fluid (see Fig. 8). Light from an underrun incandescent bulb is focussed and directed at the face of the prism contacting the liquid. The angle at which the beam hits the prism can either be cyclically varied using a rotating spiral slit, or the change in the position of the refracted beam edge monitored by a large area photo detector. The prism is thermally isolated to keep it at precisely the same temperature as the process fluid, and the temperature is accurately

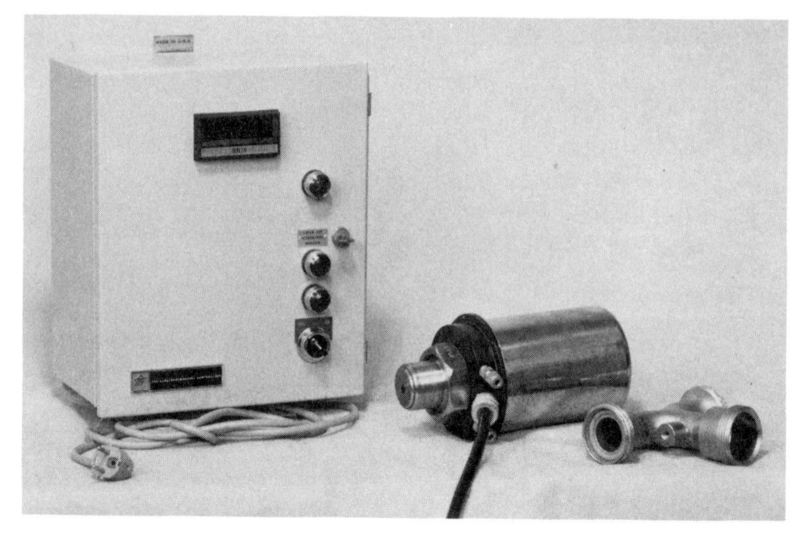

**Fig. 8.** Process refractometer—head and read-out unit (courtesy Moisture Systems Ltd).

monitored by an adjacent platinum resistance thermometer—for example, with water a 1°C shift in temperature causes a shift in refractive index of 0.00009 at 20°C: for ethanol the shift is 0.0004. Typical measurement accuracy of the refractive index is 0.0001, with precision quoted at 0.00003. Spans can range from 0.0015 to 0.130, with refractive indices ranging from 1.25 to 1.65. The performance is unaffected by colour, opacity, viscosity, flow rate, entrained gas or varying pressure. However, the phenomenon used to measure the refractive index is a surface one, so that it is essential to keep the prism surface in contact with the liquid clean. Various methods are available from manufacturers, including rubber wipers, ultrasonic probes, solvents and, for difficult cases, steam jets. In spite of this limitation, process refractometers have been successfully used with some very dirty process streams.

The advantages of this method of determining water are:

(a) the sensors are precise, reliable, ruggedly made and are available in explosion-proof form
(b) measurement is free from effects from other physical properties of the liquid except temperature, which can be compensated for automatically
(c) it is a true in-stream device, with automatic prism cleaning attachments available.

The disadvantages are:

(a) as with all physical property sensors, the system must behave as a binary one as far as the parameter measured goes
(b) it can be affected by deposits on the prism face which are resistant to existing cleaning methods
(c) it is relatively expensive

Applications include the determination of water in fruit juice, soft drinks, wine, sugar solutions, wort, acid, soap solutions, caustic soda and paper-making by-products

### 3.2.6 Other Physical Property Sensors

As must be stressed again, these all have one factor in common—they can determine water only in systems which can be considered as consisting of two components—water and one other of fixed physical property. Conductivity is widely used, both with electrodes and in the electrodeless form, where the liquid essentially acts as the core of the transformer or choke. The latter technique is of especial value in determining the water content of highly corrosive liquids such as mineral acids. The electrode-using sensors can have very low limits of detection when determining traces of water in non-conducting solution (sufficient ions have to be present to render the solution conductive) and can at the other extreme be used to determine the purity of distilled and deionised water. Temperature compensation is essential and is normally made automatically.

Dielectric constant measurements are simple and non-intrusive, but are affected by the presence of entrained gas and electrolytes, and have their main field of application with solids (see Section 5.2.5). Density can be determined with high precision on stream and has been widely used to determine the water content of slurries. The attenuation of $\gamma$ rays—usually from $^{137}Cs$ or $^{60}Co$ with larger pipes, $^{241}Am$ for narrower ones—and the vibrating tube technique are the most widely used. In the latter case the resonance frequency of a pipe conveying the liquid is continuously monitored. This frequency depends upon the mass of liquid in the tube and hence its density. Entrained gases affect both techniques of density measurement (indeed some vibrating tube gauges cease to function above a certain level of entrained gas) although there is a technique for overcoming this 'gas error' on line[4]. Viscosity can be a very sensitive indication of water content since it is highly dependent upon it in a number of cases. However, the materials have to be very well defined in terms of its other physical properties—especially temperature—dispersion of any solids present and even its recent history.

In general the advantages of physical property sensors are:

(a) they are simple, easily applied instruments
(b) they are often non-intrusive.

Their disadvantages are:

(a) the system analysed must be able to be considered binary in composition
(b) entrained gases often interfere
(c) variations in other physical properties, especially temperature, require compensation
(d) normally they are only suitable for major amounts of water (>5–10 per cent)

The range of application is very wide, within their quoted limitations. Because their performance is very application-dependent, they tend to fill particular niches in the measurement world rather than being of general use in the determination of water in liquids. One area of general application is in the control of dilution of base stocks in soft drinks and related industries.

### 3.2.7 Other Sensors for Determining Water in Liquids

On-line mass spectrometry and gas chromatography can be used to determine moisture in liquid which can readily be vaporised—but they would normally only be used when water is required as one part of a multicomponent analysis. They are sensitive—limits of detection 10 ppm are possible—and the mass spectrometric technique is very fast. Both can present problems in terms of water adsorption and materials of construction. Microwater attenuation, largely used for determining moisture in solids (see Section 5.2.4) has also been used to determine moisture in liquids such as crude oil, although the majority of so-called liquid applications are actually emulsions or pastes, such as are common in the food industry.

## REFERENCES

1. Pitt, G. D., *Electrical Communication*, **57**(2) (1982) 102–6.
2. Cornish, D. C,. Jepson G. and Smurthwaite M. J., *Sampling Systems for Process Analysers*, Butterworths, London, 1981, pp. 25–95.
3. Willis H. A., *Advances in Infrared and Raman Spectroscopy—2*, Heyden, London, 1976, pp. 111–12.
4. UK Patent Application No 8021717, 1981.

CHAPTER 4

# *On-line Moisture Measurement in Slurries, Pastes and Emulsions*

The on-line determination of moisture in slurries, pastes and emulsions is of great industrial importance but can present considerable problems. The materials possess varying degrees of heterogeneity which complicates sampling and sample presentation. In addition they may be difficult to convey, rapidly separating and highly abrasive. Water content may be high and, where, for example, viscosity is being inferred for control purposes from the water content, the latter may need determining with high precision. Because free water is almost invariably present, techniques such as temperature difference and equilibrium relative humidity are not of great value with these materials, with the exception of lower water content pastes.

## 4.1 SAMPLING SYSTEMS

The difficulties in reliably and representatively sampling these materials tend to strongly favour techniques such as infra-red reflectance and $\gamma$ ray attenuation, which can be used directly on the process line or vessel. However, if sampling is essential there are certain general observations which are of value. Firstly, a representative sample must be obtained. This implies that the material is adequately mixed at the point of sampling—hence premixing with a static mixer, or with slurries and suspensions, ensuring high turbidity at the sampler are essential. Isokinetic sampling is often stressed as being essential to obtain a representative size distribution of particles in gas/particulate streams (this technique ensures that the velocity in the sampling probe is the same in magnitude and direction as in the main stream), but in the author's experience is rarely worth the added complexity with liquid/particulate streams. With slurries sampling can be achieved by taking the flow out of a highly turbid constant head tank, by sampling on a side pipe at the output of a centrifugal pump or by inserting a circular or slit-shaped probe across the stream into a pipe or launder where turbulent flow occurs. With pastes the material handling problems may dominate good sampling practice—reliably splitting a stream into major and minor flows without suffering regular blockage may be all that can reasonably be expected. With all these materials cross cutting a freely falling stream provides the best approach to representative sampling, if it can be achieved—the sampled cross cuts being fed to a holding tank or hopper and then allowed to flow or be pumped to the sensing system. All materials and components in any sampling system have to be designed to minimise the effects of abrasion, and the whole system has to be designed, if possible, to empty and to be easily cleanable in the event of pump or power failure. A typical slurry system is shown in Fig. 9.

It is not possible in a book of this nature to discuss the fundamental and complex problems involved in devising sampling systems for slurries, pastes and emulsions. The reader is referred to standard works on the

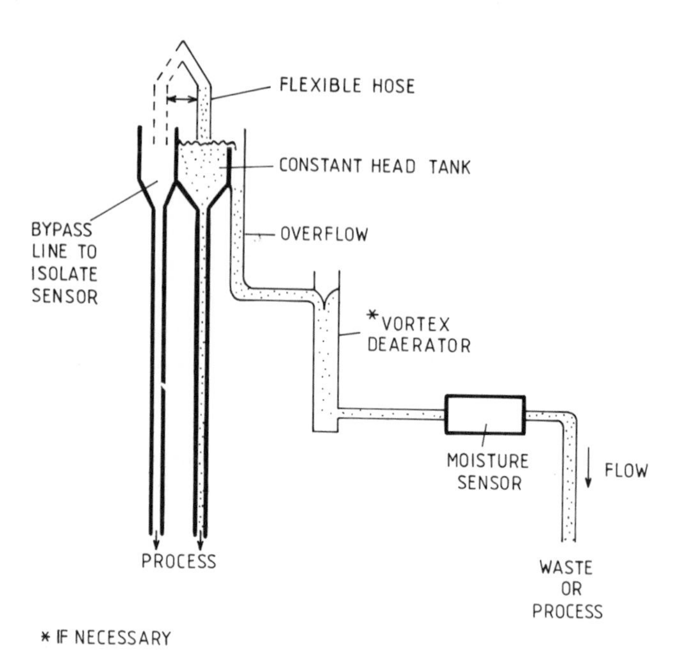


**Fig. 9.** Typical slurry sampling.

subject[1], and is recommended to seek expert advice if initial trials of a basic system indicate problems.

## 4.2 SENSOR TYPES

These materials fall between solids and liquids in their properties. Hence, sensors used for both these classes of material may be applicable and are fully described in the appropriate chapters. This chapter will highlight particularly relevant aspects of the most suited techniques.

### 4.2.1 Infra-red Reflectance

This may be used directly on a process stream (see Section 5.2.6), either on a surface exposed on a belt or tank or through a suitable infra-red transmitting window—hence it avoids the use of sampling systems. It is a surface technique, with penetration depths of a fraction of a millimetre or less, so it is essential that the surface looked at is representative of the bulk material—therefore settling or surface 'wetting' must be avoided. It also gives poorer performance at very high water contents, and thus is less applicable to liquid slurries and suspensions. Problems can occur if the surface of the material shows specular, i.e. mirror-like reflection, since this results in a very high reflected intensity, which saturates the detector and measurement system. The remedy is to break up the surface by agitation or by using a geometry that rejects the reflected radiation, although this may not work with some materials.

Its advantages for this class of materials are:

(a) it is non-contacting
(b) it can give quantitative results with irregular surfaces
(c) it is selective and rapid.

Its disadvantages are:

(a) it is a surface technique, hence the surface must be representative of the bulk of the material—or reproducibly non-representative if all else fails
(b) it can have limited accuracy at high water contents
(c) specular reflection can cause problems.

It has been used to determine the moisture content of a wide range of slurries and pastes, including margarine, cottage cheese, paper pulp, coal slurry, sewage sludge, clay, cake mixes and paste based animal foods.

### 4.2.2 Capacitance Sensors

While capacitance techniques (see Section 5.2.5) are normally contacting, the electrodes can be placed in the walls of pipes and the bases of chutes and belts, and hence the technique is applicable to this range of materials. However, most gases, including air, have very low dielectric constants compared with solids, and if entrained in the material, cause major errors. Since it may be difficult to prevent this, particularly with the thicker pastes and suspensions, this is a real limitation in its use. Its relatively poor limit of detection is not a limitation with this class of materials, since water is normally present as a major component. Correction for entrained gas may be possible with slurries by measuring the capacitance at two different pressures.

Its specific advantages for these materials are:

(a) its electrode can be made in a number of forms to fit appropriate conveyors and ducts, and of materials having high abrasive resistance
(b) it is a good technique for determining relatively high water contents.

Its disadvantages are:

(a) it is bulk-density and included-gas sensitive
(b) it is normally a contacting method.

It has been used to determine the moisture content of butter, foundry pastes and sugar beet pulps.

### 4.2.3 Low Resolution Nuclear Magnetic Resonance (NMR)

This technique, fully described in Section 5.2.7, determines the concentration of hydrogen atoms in the liquid phase—and hence the water content. It is not contacting, but because it determines the mass of protons present in a restricted volume, the bulk density of the material in that volume must remain constant or be independently measured. Hence, entrained gases cause errors. As the material has to pass through an intense magnetic field during measurement, the technique is largely restricted to small bore ($<50$ mm) pipes, which places limits on the rheological properties of the materials that can be examined.

Its advantages when used with these materials are that it is non-contacting and proton selective. Its disadvantages are:

(a) it is restricted to small bore pipes
(b) it is bulk-density—and hence entrained-gas—sensitive.

It has been used to determine water in fats and other paste like food intermediates and products.

### 4.2.4 Microwave Attenuation

This technique measures the attenuation of microwaves across a fixed path (see Section 5.2.4), which can be used to determine the moisture content since water strongly absorbs the radiation. The measurement does require a path free of metallic conductors, which means either using PVC ducting, for example or having windows of ceramic or plastic placed in the zone where measurement is required. It can be used either with a transmission path across the sample, or with strip line techniques, which essentially measure from one side. The method, like capacitance measurement is bulk-density dependent, although phase measurement methods exist for minimising this effect over limited density ranges, without the need to independently measure bulk density[2]. The instrumentation required is more expensive than for capacitive measurements, but is less sensitive to effects due to electrolytic conduction and to variations in chemical composition.

Its advantages for pastes, slurries and suspensions are:

(a) it is non-contacting
(b) it measures up to relatively high water content
(c) it is relatively insensitive to the effect of electrolytic conductivity
(d) at its lower frequencies it is a highly penetrating technique.

Its disadvantages are:

(a) it requires a non-metallic path
(b) it is bulk-density dependent in current commercial models.

It has been applied in the food industry to suspensions and pastes. With the availability of newer ways of applying the technique, particularly those measuring reflected rather than transmitted power, the range of applications in these materials is likely to increase.

### 4.2.5 On-line Viscometers

On-line viscometers have been used to determine the water content of slurries, pastes and suspensions[3]. Over certain ranges the viscosity will be highly dependent on the water content.

These materials rarely show simple Newtonian viscosity, the apparent viscosity being highly dependent on shear rate. Readers are directed to standard works on this highly complex subject, but it should be pointed out that the viscosity depends on a number of other physical characteristics, such as particle size, particle shape, surfactants and temperature. Hence, to be a useful means of determining water content, all these variables must be closely controlled.

On-line viscometers fall into three main types:

*(a) Rotational*

These use an element, of a variety of shapes, rotating in the material being characterised. The shear stress is measured from the rotation torque imposed on the measuring element by the material. For non-Newtonian fluids, the shear rate depends on the non-Newtonian characteristics; and techniques are available to determine the correct shear rate for measurement purposes.

*(b) Tube*

These measure flow rate and pressure drop through a tube and derive a value of the viscosity from known relationships. Here again the non-Newtonian characteristics must be allowed for in their application. The tube can vary between a capillary (obviously not suitable for the majority of materials in this chapter category) to pipes of several centimetres diameter. Where the velocity is constant—for example if the material is being propelled by a ram at controlled velocity—this is a simple method of measurement, essentially requiring a single pressure sensor.

*(c) Falling body*

These measure the steady rate of drop of a body through the fluid, viscosity being derived using Stokes' Law. The calculation becomes very complex for non-Newtonian liquids, and the on-line sensors using this technique normally use an empirical derivation.

The main advantages of process viscometers for water determination in slurries, pastes and suspensions are:

(a) they can show good sensitivity, i.e. change in signal with water content
(b) the measurement can be very simple.

The disadvantages are:

(a) the viscosity depends on a large number of other parameters
(b) most of these materials are non-Newtonian, which complicates viscosity measurement.

The classic use of these measurements for determining water content was in the brick industry, where the pressure before the extruder was used to determine the moisture content of the clay. Other applications include water in paper-making pulp, and water addition to coatings and surface finishes.

### 4.2.6 Other Physical Property Sensors

A wide variety of physical property sensors has been used to determine the solid, and hence the water, content of slurries, pastes and suspensions. Density gauges are used to determine the water content of slurries—particularly the $\gamma$ ray attenuation gauge in the mineral processing industry, for it can be made in a highly rugged, non-intrusive form for large pipes. Ultrasonic attenuation has been used with sewage sludges and mineral slurries—but interference due to scattering from suspended air must be removed, since an air bubble will scatter as well as a solid particle. Light scattering has been used for lower solid contents, but with the same constraint. Where these techniques are used, it is generally because of the simplicity of the measurement, or because their in-stream applicability eases the problems inherent in the representative and reliable sampling of slurries, pastes and emulsions.

## REFERENCES

1. Cornish, D. C., Jepson, G. and Smurthwaite, M. J., *Sampling Systems for Process Analysers*, Butterworths, London, 1981, pp. 96–258.
2. Meyer, W. and Schilz, W., *J. Phys.*, *D.* Appl. Phys., **13** (1980) 1823–30.
3. Tily, P. J., *Measurement and Control*, **16** (1983) 111–15, 137–39.

# CHAPTER 5

# *On-line Moisture Measurement in Solids*

Moisture measurement may be required on solids ranging from fine free-flowing powders, to rock emerging from a jaw crusher. It may be present in the rotating drum of a dryer, on a conveyor belt, as a continuously moving web or in a covered stockpile. This complexity immediately points to the first major problem in moisture determination in solids—ensuring that the sensor sees or penetrates a representative sample of the material.

## 5.1 SAMPLING SYSTEMS

The first rule with solid sampling systems is 'if you can otherwise obtain adequate information on water content for control purposes, avoid them'. This puts systems such as infra-red diffuse reflectance, capacitance and microwave attenuation at a premium, since these can be used directly on the sample stream or in the containing vessel. However, there are circumstances—for example very low moisture contents or very large scale flows—where sampling becomes necessary. The subject is, as with slurries, pastes or suspensions, very complex, and the reader is referred to standard works on the subject[1]; however there are some basic guidelines which are of value.

Normally in solid materials, there exists a range of particle sizes—and chemical composition may vary with the particle size. Water present may also not be uniformly distributed with particle size—for example if it is mainly retained on the surface, the large particles, having a smaller surface area for a given mass, will have a lower moisture content. Thus any sampling system must ensure that an adequately representative range of particle sizes is taken. 'Adequately representative' needs a little amplification. If for example the material contains occasional large lumps, which make up only 1 per cent by weight of the material, then, unless their water concentration is vastly different than in the remainder—a very unlikely situation— they can just be rejected and do not need including in the sampled material. This would not be representative, but would give a good indication of moisture content for control purposes.

Representative sampling of solid materials—particularly coarser ones— is best achieved when the material is in free fall—say from the end of a conveyor belt. A suitable vessel can be passed across the falling stream at regular intervals to take an incremental sample, which can be retained in a hopper and either then subsampled after grinding or analysed directly. The whole of the design of such systems[2] is well established, but they do require high quality engineering design to make them reliable, especially with larger scale installations.

With finer materials, there is a much wider range of devices which give adequately representative and reliable sampling of solids, their design often being dictated by the type of material and the siting of the sampling point. Material on a conveyor belt can be intermittently sampled by a scoop that crosses the belt. Powder in an air slide or similar gas/solids conveyor can be blown out by the pressure in the slide through a slit-shaped or circular tube pointing upward towards the flow in a position away from the slide wall. Where materials, such as grain, have a relatively regular shape and particle size and where moisture is regularly distributed, sampling becomes mainly the mechanical problem of regularly and reliably removing a suitably sized sample, and simple double headed valves working through the side of a chute or slide-like automated collectors can be used (see Figs 10 and 11). Actuation of all these systems is

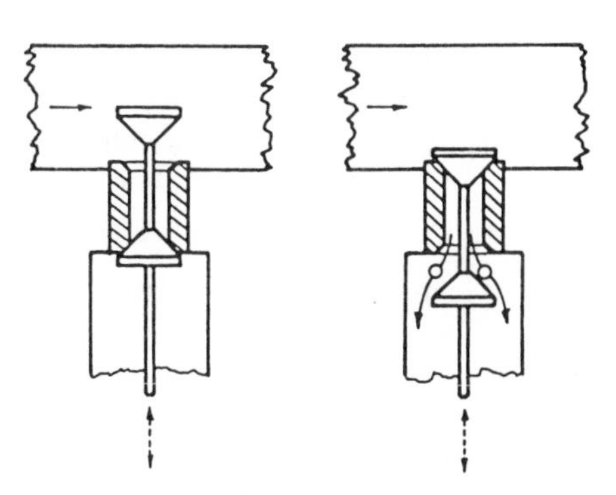

**Fig. 10.** Poppet valve solid sampler.

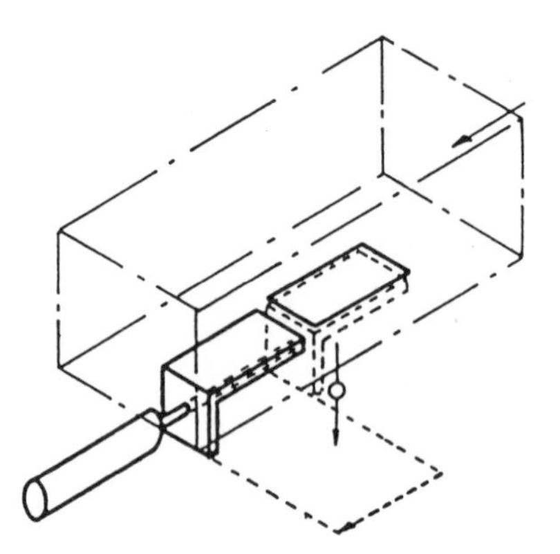

**Fig. 11.** Slide or gate solid sampler.

preferably pneumatic since such means are capable of high power and are intrinsically safe— which is important when the possibility of dust explosions exists.

Even where on-stream sensing techniques are used, some simple device may be necessary to assist the moisture determination. When infra-red reflectance analysis is carried out on belt-carried materials, a simple rake preceding the sensor produces a fairly regular and fresh surface for analysis. Microwave techniques require a non-metallic path, and a PVC or other plastic chute with a means of holding the flowing material at roughly constant bulk density may be used.

## 5.2 SENSOR TYPES

As with the rest of the book, these are described by the techniques employed. The use of sensors to determine the form in which moisture is present in materials is the subject of a separate chapter.

### 5.2.1 Loss in Weight

The moisture content of a material can be determined by loss of weight on drying at a controlled temperature—one of the classical techniques of analysis. It can be interfered with by loss of other volatile components or retention of moisture inside larger or moisture-impermeable particles. While it is possible to automate and place on line with a sampling system the laboratory-scale loss in weight techniques—now accelerated by the use of microwave heating and microprocessor based tareing and calculation—such a system would be mechanically complex and the author knows of no industrial applications for control purposes. Where the reactor or dryer is mounted on load cells, loss in weight can be used to determine the moisture loss, as could belt weighing before and after an industrial-scale infra-red or microwave dryer. The rate of change of weight at a controlled temperature can be used for dryer control, but it must be stressed again that the possibility exists of positive or negative bias for the reasons given above.

The advantages of this technique are:

(a) it employs an absolute measurement
(b) the measurement can be directly made in the process vessel.

The disadvantages are:

(a) it is mechanically complex on a smaller scale
(b) it is subject to possible negative or positive bias due to loss of volatiles or particle-trapped water.

### 5.2.2 Mechanical Properties

Some mechanical properties of powders, such as cohesiveness and resistance to flow, are highly dependent on the moisture content, and a simple means to measure the former can be of value in estimating the latter. One of the classic applications—still in use by at least one UK pharmaceutical manufacturer—is the taking of a handful of product from a dryer and seeing if it makes a cohesive ball after squeezing in the hand. A somewhat more quantitative application is the determination of moisture in sand. Here the ability of a sand bed to pass across slots of varying width can be correlated with its water content. Such applications are highly specific for the particular problem, but can be simple and reliable, although a certain amount of ingenuity may be required to interface the measurement to on-line control systems.

Their advantages are:

(a) they can be simple and reliable
(b) often they are low cost.

Their disadvantages are:

(a) they are highly application-dependent
(b) they are often difficult to interface with microcomputer control systems.

### 5.2.3 Temperature Difference

When a water-containing solid is being dried by passing a gas over it, a temperature difference will exist between the material and the drying gas or the ingoing and outgoing gas streams, until the vapour pressure from the water in the solid reaches equilibrium with that in the drying gas. The temperature drop is caused by the requirement of the latent heat of vaporisation of the water. Normally drying is not carried on to the stage where there is no significant difference between the two temperatures. In many dryers empirical selection of a certain temperature difference can ensure good drying control, either as the end point for a batch, or to adjust the feed of material through a continuous dryer to maintain the selected temperature difference at the discharge end of the dryer. It is a simple, inexpensive and reliable method for dryer control[3]. Problems found in applying the technique include difficulties in measuring the true solid temperature—sensor to material contact may be difficult to maintain consistently—and the effect of water gradients within the particles being dried. The latter are nearly always present in particles as they dry, and they have probably been the cause of failures in the application of the temperature difference technique.

Reasons for adopting the technique are:

(a) the equipment is simple and reliable
(b) the low cost
(c) it is reasonably easy to install.

Arguments against are:

(a) it fails with a number of materials
(b) it is empirically calibrated when a fixed temperature difference is used.
(c) it has poor limits of detection.

It has been successfully applied to spray and fluid bed dryers in the pharmaceutical and food industries, and to the drying of textiles.

### 5.2.4 Microwave Attenuation

Water strongly absorbs microwave radiation at certain frequency bands (typically two to three orders of magnitude more strongly than the base material), and this has been used to determine the moisture content of a wide range of powders and granular materials[4]. One of two frequency bands is normally used—the S band (approximately 1–2 GHz) or the X band (approximately 9–10 GHz). The S band is more penetrating and hence is better suited to the larger scale, more rugged on-line determinations, for example in the cement and steel industries. The X band, while being less penetrating, shows reduced sensitivity to the composition of the material being measured and is the most widely applied at the present time. The sensing system is basically simple, consisting of a microwave power generator, wave guide to and from the sample cell, detector—normally, although not necessarily on the side of the sample cell remote from the source—and a power meter with suitable output for a control system. Horns are used between wave guides and cell to reduce reflection from the junction between cell wall and sample. Where more appropriate for the type of sample stream, strip line techniques of measuring the attenuation can be used.[5] Modern microwave generators and detectors are all solid state and are capable of showing a high level of reliability.

Typically, a Gunn diode is used as a source, and a Schottky diode as a detector. Powers used are very low, 0.5–1.0 mW $cm^{-2}$, so that no known health hazard is involved.

The output of most commercial instruments is density-dependent, and if the bulk density of the material under examination varies to a degree greater than the maximum tolerable error, a $\gamma$ ray attenuation or other bulk density meter is included to allow correction for these variations. However, with many materials such compensation can be avoided by the careful selection of flow cell and material feeding mechanism. For example, feeding of grain into a vibrating non-metallic hopper, through which the microwave radiation is passed, has been shown to give good on-line results. Attenuation is also temperature-sensitive, and if this varies significantly, then measurement and compensation must be used. Methods have been developed which, by measuring the real and imaginary components of the radiation simultaneously in the X band[6], give an output which is essentially independent of bulk density over limited ranges of water content—a situation often encountered in process control, where the highest precision is required near the set point. Such systems are now becoming commercially available and should greatly extend the value of the technique.

Limits of detection are in the range 0.3–0.5 per cent water, with normal operating ranges between approximately 1 and 70 per cent water.

Transmission is generally used for lower water concentrations with reflection being measured at higher ones.

Accuracies of $\pm 0.5$ per cent of water are quoted as being achievable over most of the range.

The advantages of the technique are:

(a) physical contact between the sensor and material is not essential, although care must be taken to ensure good contact in order to minimise reflection losses at the interfaces between sensor and sample
(b) the material is not contaminated nor altered in composition
(c) the measurement system, being low power and all solid state can be proofed against the ingress of dust
(d) the microwave powers used are very low and are considered to be completely safe
(e) the microwaves pass through the bulk of the material and hence give more representative results than surface-sensitive techniques where inhomogeneity, such as moisture gradients, are present.

The disadvantages of the technique are:

(a) relatively poor limits of detection—between 0.3 and 0.5 per cent at best
(b) the need, if lower moisture ranges are of interest, of transmitting through a bed of material which is of limited thickness—typically 5–12 cm
(c) dependence, for most current sensors, on the bulk density of the material being analysed
(d) calibrations are composition-dependent—differing materials giving separate calibration graphs
(e) the need to have a non-metallic path between source and detectors
(f) effects due to standing waves caused by impedance mismatch.

Microwave moisture measurements are becoming more popular for on-line applications, due in part to the availability of reliable all solid state microwave sources and detectors, in part to the use of the 10 GHz spectral region, which is less dependent on composition effects, and also to the use of microprocessors to carry out such functions as power monitoring, linearing and bulk density compensation. They have been used on-line for the determination of moisture in paper, flour, grain, coal, limestone and cement. Generally the technique should be considered for materials that do not vary widely in composition and can be readily handled, such as powders or granules that are free flowing. The technique is less suited to determining moisture in reactors or in other plant where the chemical composition can vary independently of the moisture content.

### 5.2.5 Capacitance Measurement

The dielectric constant of water is considerably higher than that of most other materials and this factor is used to determine the moisture content of solids such as natural vegetable products, which do not vary widely in composition[7]. The dielectric constant also depends on the bulk density and chemical composition of the solids—especially varying concentrations of ionic conductors such as salt—and it is for this reason that the method is mainly of value with materials of roughly constant composition. However, in the process control field this may be less of a limitation than at first appears, since many raw materials, intermediates and final products fulfil this qualification. The capacitance is normally measured at radio frequencies (2–12 MHz being typical), the material being allowed to flow at an adequately constant bulk density between two metallic plates which make up the capacitor. The plates can be hard surfaced with, for example, chromium or a non-conducting ceramic if wear or corrosion demand it. The capacitance can also be measured between two plates placed side by side, giving a 'from one side' measurement of especial value with conveyor belts or slides. The capacitance is determined in the majority of instruments by high precision bridge techniques, with built-in compensation for variations in material temperature (an equivalent of 0.05 per cent moisture per degree centigrade is a typical effect), variations in bulk density and the effect of electrolytic conduction. Power loss measurement techniques are also employed. As the electrodes of the capacitor can be very rugged, operation at high temperatures is possible and direct in-kiln or in-dryer determinations are possible at temperatures up to or above 250°C.

In suitable applications the method has a limit of detection of 0.1 per cent moisture and repeatability of ±0.2 per cent moisture over the range up to 30 per cent moisture. It can be used with somewhat poorer performance up to about 60 per cent moisture where repeatability can range between 0.5 and 2 per cent moisture.

The advantages in using this method are:

(a) it is a simple, semi-invasive technique which is readily installed
(b) it is relatively inexpensive and reliable
(c) in suitable applications it has a reasonably good limit of detection and repeatability
(d) it can be made very rugged.

The disadvantages of using it are:

(a) it is bulk-density, temperature and chemical composition dependent

(b) it requires electrically isolated conducting electrodes very close to the material being sensed.

It has been used to determine the moisture content of a wide range of foodstuffs, tobacco, chemicals, sugar beet pulp, fertilisers, drugs, soap flakes, powdered coal, sands, wool and textiles.

### 5.2.6 Infra-red Reflectance

All forms of water strongly and selectively absorb infra-red radiation[8]. The wavelength bands normally employed when the water is in liquid phase centre on 1.45, 1.94 and 2.95 $\mu$m although there are weaker ones at 0.76, 0.97 and 1.18 $\mu$m, and a strong one at 6.1 $\mu$m which is outside the normal wavelength operating range of this technique. Water vapour also absorbs, but at somewhat shifted wavelengths, with its infra-red spectrum showing the mass of narrow absorption lines associated with gaseous spectra. This shift and change in the nature of the gaseous spectrum allows infra-red reflectance determinations of water in solid materials to be carried out even in the presence of water vapour in the infra-red beam to and from the sensing head. Thus the majority of applications can be carried out with a normal air path between the sensing head and the material being analysed—a major advantage for the technique. The strengths of the three normally employed absorption bands quoted above increase with wavelength—the one at 1.94 $\mu$m being the one most widely applied in on-line moisture determination.

Powdered, granular and coarser solids are generally opaque to infra-red radiation, it being rapidly attenuated by a combination of scattering at the air–solid interfaces and absorption within the material. Path lengths of a fraction of a millimetre would be required to obtain measurable transmitted intensity, which is clearly extremely difficult in an on-line situation. Not only that, but variations in transmission due to changes in scattering conditions such as bulk density would be very large. Therefore, diffuse reflectance from the surface of the material rather than a transmittance method is used for the on-line determination of moisture in solids. In a typical sensor, infra-red radiation from a quartz iodine bulb—underrun to give long life—is directed as a parallel beam by means of a lens or mirror onto the surface of the material being analysed. The infra-red radiation is diffusely reflected from the material, but much less so at wavelengths where the material strongly absorbs. These wavelengths will include the moisture absorption bands when significant amounts are present. The diffusely reflected radiation is collected by a large aperture

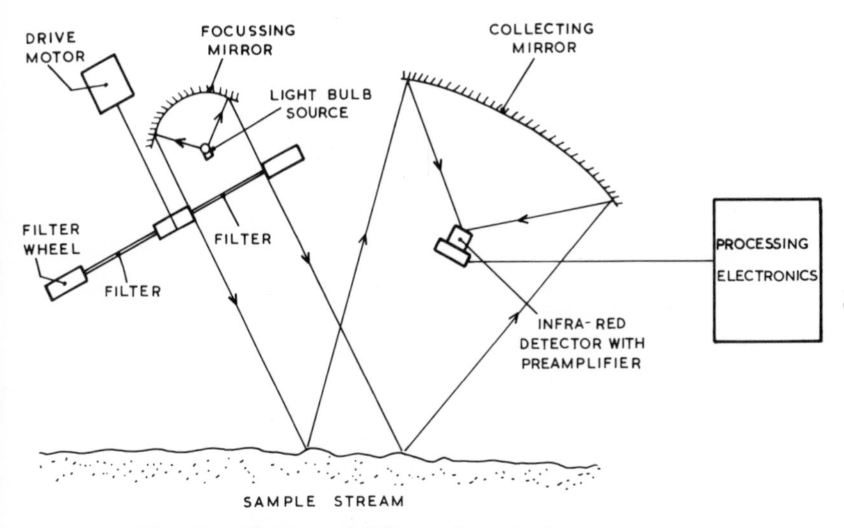


**Fig. 12.** Diagram of diffuse infra-red reflectance system.

mirror or lens and focussed onto the detector which is normally a lead sulphide cell which may be thermoelectrically cooled (Figs. 12 and 13). A wheel, containing two or more narrow band infra-red interference filters, is inserted in the beam from the source; one filter transmits infra-red radiation at a wavelength which water strongly absorbs, such as 1.94 $\mu$m, while the second and/or third transmit at wavelengths where water does not strongly absorb, for example 1.8 $\mu$m. The two or more differing intensities detected as the filters sequentially spin across the beam are sorted electronically, measured independently and then ratioed to give an output largely independent of variations in source power output, position of sample or detector sensitivity—although all these effects are still present to a limited extent. One interesting version of the instrument exists which uses infra-red light emitting diodes to provide the differing radiations required. These emit modulated monochromatic infra-red radiation, which is conveyed to the sample with fibre optic bundles. Another bundle returns the diffusely reflected radiation to the photo detector. With this system a compact, rugged sensor head with no mechanical moving parts is achieved which has the potential of being made in an intrinsically safe form. Because the diodes are currently only available in the shorter wavelength region, this sensor's application field is limited to higher moisture contents such as those found in animal feedstuffs. Where the scattering power of the material is very variable,

**Fig. 13.** Diffuse infra-red reflectance gauge (courtesy Infra Red Engineering Ltd).

the scattered intensity on the lower and higher wavelength regions on both sides of the water absorption peak are measured, and the mean of these intensities used to calculate the ratio. The logarithm of the intensity ratio is calculated, and the instrument then relates this to water content with an in-built calibration graph, specific for the material being analysed.

With powders or granular materials the surface is crudely stirred and smoothed just before analysis—for example, for materials on a conveyor belt a simple fixed rake is adequate. The method is, within normal operating limits, self-compensating for variations in bulk density. With materials falling in a duct, a probe with an infra-red transmitting window may be inserted so that the material impacts upon it—keeping it clean and presenting a continuously varying sample to the beam. Many successful applications also are found with coarser materials such as tobacco, rubbish or beet pulp. The tolerance on the sample surface position relative to the sensor is wide—a range of 50 to 300 mm can be accommodated in many applications without noticeable change in the calibration. It is this ease and flexibility in coupling sensor to sample stream which is one of the major factors making this so powerful a technique.

The sensor, as has already been pointed out, requires calibration for each application. The initial calibration may be obtained in the laboratory looking at a series of samples with established water contents. Ideally, these should be samples taken from the process with varying water concentrations, for in some applications water additions can give a somewhat different calibration. Sometimes this is all that can be achieved, but if possible the calibration should be confirmed or trimmed with a series of representative samples taken from the stream as the sensor measures its water content. Concentrations varying from 0.02 to near 100 per cent have been determined, and accuracies of $\pm 0.5$ per cent relative have been achieved although $\pm 1$ per cent relative is more typical. Both the concentration ranges and accuracies are of course application-dependent. Problems can arise when the technique is used with highly infra-red absorbing materials, such as carbon black, or those showing appreciable specular or mirror-like reflectance. In the former case the limitation is receiving insufficient energy to make a measurement. This can only be minimised by optimisation of the instrument parameters—as high an intensity source as possible, maximum aperture optics, high sensitivity detector and long integrating times in the electronics. In the latter case either too high an intensity is reflected from the specular surface—saturating the measurement system—or if the geometry can be arranged to avoid the specular beams, too little radiation penetrates the sample and the system suffers from lack of energy and sensitivity. While surface roughening may provide a solution, in some cases specular reflection makes the technique unuseable.

The reasons for using the technique are:

(a) measurement is made by a non-contacting, remote head which can be water cooled in high ambient temperatures and air flushed in dusty conditions
(b) it can measure moisture content of material in an enclosed space, through a suitable infra-red transmitting window
(c) it has a wide range of measurement, with good selectivity and high reproducibility
(d) it has a very large range of application.

The reasons against using it are:

(a) the head cannot in its conventional form be made intrinsically safe, due to the power required for the infra-red source, coolers, etc.; it can be obtained in an explosion-proof form
(b) the technique involves a surface determination and this must be representative of the bulk of the material being analysed
(c) it is somewhat more expensive than some of the simpler techniques
(d) problems can arise with highly absorbing materials, such as carbon black, or materials giving a high mirror-like or specular reflection.

Applications include the determination of moisture in sands, clays, chemicals, grains, foodstuffs, concrete, wood shavings, tobacco, soap fertilisers, textile webs, paper, powdered coal and iron ore sinters in belts, dryers, mixers and reactors. It is well suited to determining the end-point in dryers, and has a good reputation for reliability with industrial users.

### 5.2.7 Low Resolution Nuclear Magnetic Resonance (NMR)

All hydrogen atoms have a nuclear magnetic moment—that is they behave like extremely small bar magnets. The laws of quantum mechanics allow these magnets to align themselves in any magnetic field only at certain angles to that field. Corresponding energy quanta are required to shift the atoms from one alignment angle to another, and the size of these quanta depends on the strength of the magnetic field. When hydrogen atoms in a magnetic field are exposed to electro-magnetic radiation at a frequency corresponding to the energy quantum required to move the atoms from one alignment to another, then selective resonance absorption of this radiation occurs. This applies to all hydrogen atoms, irrespective of what compound they are present in, since it is a nuclear effect. The molecular environment does cause small shifts which can be measured with high resolution nuclear magnetic resonance spectroscopy—an important technique for determining molecular structure—but these do not affect the low resolution measurements significantly. Of more importance is the phase in which the molecules are present. Molecules in a liquid phase give much sharper absorption maxima than those in solids and this allows the determination of hydrogen-containing liquids, such as water[9], in hydrogen-containing solids, such as coal or even butter. The time taken for the hydrogen atoms to align themselves when placed in a magnetic field—or to randomise their orientation when released from it (called the relaxation time)—can be used to indicate the form in which the water is present; this is further discussed in Chapter 6. The technique is normally abbreviated as NMR.

A typical on-line instrument contains a sensor head (see Figs 14 and 15) with a remote read-out unit. The sensor head consists of a permanent magnet with a gap for the sample stream, sweep coils to modulate the magnetic field over a small range (about 10 gauss or less in a typical fixed field of 600 gauss), a coil to generate and detect losses in the radio frequency power, and temperature sensors to monitor stream and ambient conditions. The read-out and control unit varies the magnetic field scanning range, monitors and compensates for peak shifts and converts the loss of signal power into proton or moisture concentration with calibration data which are material-dependent. Water can be determined over the range from 0.05 to near 100 per cent, although the actual range with any material may be less than this. An accuracy of better than 0.5 per cent relative can be achieved at higher concentrations. The technique

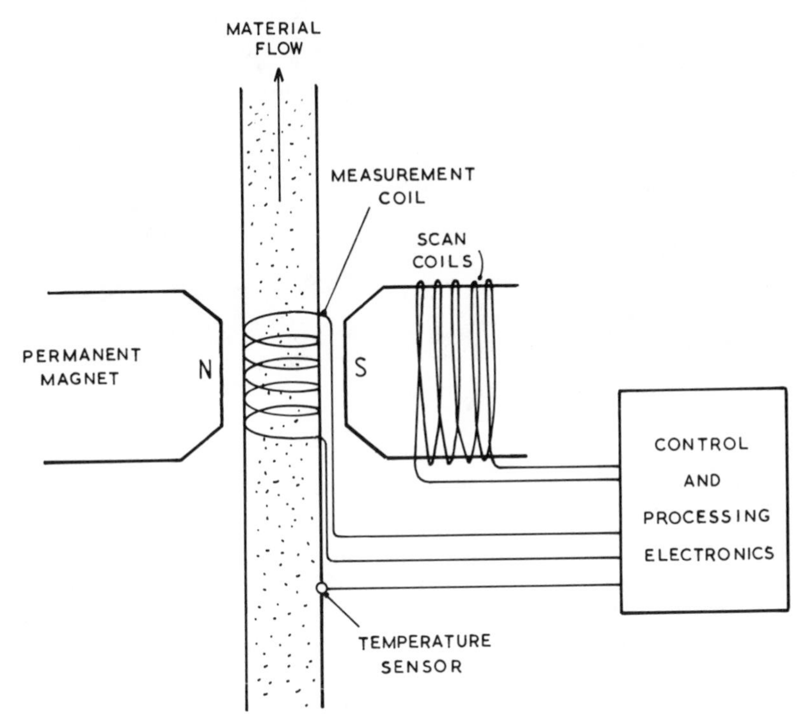


**Fig. 14.** Diagram of low resolution NMR head.

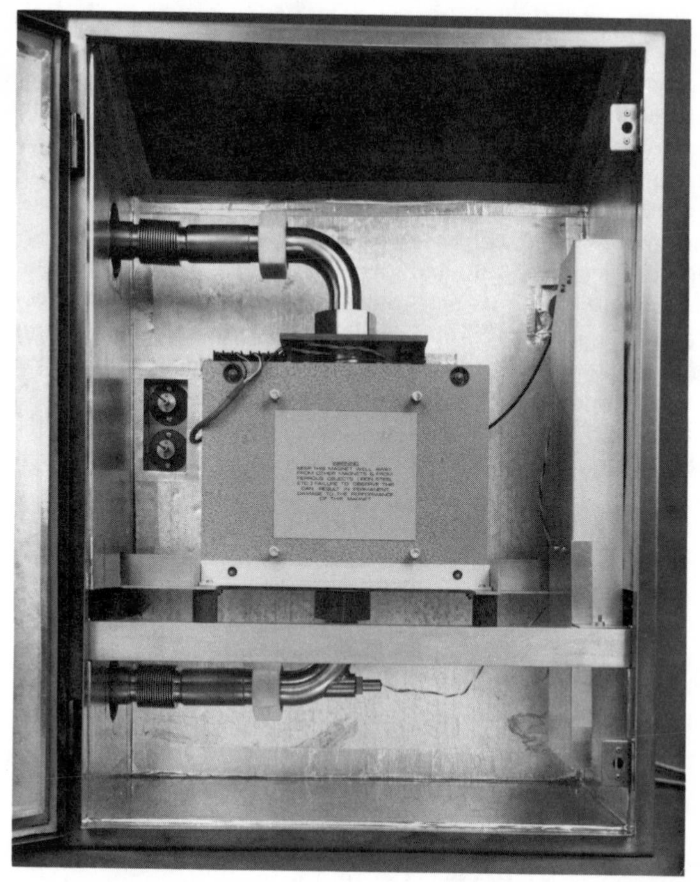

**Fig. 15.** Low resolution NMR head (courtesy Oxford Instruments Ltd).

cannot be used with materials containing significant amounts of ferromagnetic materials such as some metals and ores.

When used on-line with powders or granular materials, a sample is taken and conveyed into the gap, or a stopped flow method can also be employed. A pipe or duct of up to 50 mm diameter can be accommodated in the standard magnet, and large gaps can be used if the application demands it. The technique is flow ratio dependent with moving materials—indeed it forms the basis of one type of flow meter. Flow dependence can be reduced by having a secondary magnet upstream of the measuring magnet or making the measurement magnet longer in the direction of flow so that the material has a longer residence time in the magnetic field before the resonance absorption is measured, thus allowing the atoms to fully align to the magnetic field. It can be made unimportant if a constant flow rate is achieved, and while this is possible with liquids, greater difficulty exists with powders and pellets, thus accounting for the adoption of sample or 'stopped flow' methods with these applications. Since the technique determines the total mass of hydrogen atoms in the sensed part of the magnetic field, it is necessary either to know the bulk density of the material by measuring it on-line, or to be able to control it in the sensing zone. The technique is also temperature-dependent with a temperature coefficient in the region of $\pm 1$ per cent relative per degree centigrade.

The advantages of the technique are:

(a) it is specific for hydrogen atoms in liquids
(b) sample preparation is not required
(c) it is simple to apply
(d) it is reasonably sensitive with good potential accuracy.

The disadvantages of the technique are:

(a) it will measure hydrogen in other liquids such as oils which may be present in the sample—hence they interfere
(b) magnetic materials cause problems
(c) as it is flow- and temperature-dependent, these must be controlled or compensated for
(d) it uses a heavy magnet (45 kg) in the sensing head.

The technique has been used to determine moisture in grain, foodstuffs, building materials, cement slurries, coal and textiles. It can give information on the form the water is present in—see Chapter 6.

### 5.2.8 Neutron Moderation

This technique of moisture measurement, like NMR, determines the hydrogen concentration in the material, from which the moisture content is inferred[10]. 'Fast' (i.e. energetic) neutrons are selectively slowed by certain nuclei, of which the most abundant is the proton—the nucleus of the hydrogen atom. Thus the intensity of 'slow' neutrons around a 'fast' nuclear source is highly dependent on the concentration of hydrogen atoms in the vicinity of the source.

A typical sensor head contains a suitable source of 'fast' neutrons (such as a mixture of the $\alpha$-emitter americium-241 with beryllium), a selective detector of slow neutrons, and a secondary $\gamma$ ray sensor to compensate for variations in the density of the sample. The neutrons are highly penetrating and the sensed volume extends between 20 and 40 cm from the source, depending on the material being analysed. This penetration is of especial value with heterogeneous materials, where the moisture is non-uniformly distributed and where extreme ruggedness is required, since the sensor walls can be thick and capable of coping with the most adverse environments such as steel or cement works. The source is doubly encapsulated in arc welded stainless steel and a similarly encapsulated caesium-137 source is used to provide $\gamma$ rays for the density compensation which is usually required. A single glass scintillator, based on lithium-6 is used in the scintillation counter. It can simultaneously measure the 'slow' neutron and density-dependent back-scattered $\gamma$ rays, using pulse height discrimination. The remote read-out unit contains amplifier, pulse height selector, detector high voltage supplies and a microprocessor to check on sensor operation, compensate for source decay and last (but not least) calculate the moisture content, assuming all hydrogen present is in the form of water.

Concentrations of water from 0.5 per cent upwards can be determined with a precision at lower levels of $\pm 0.1$ per cent water, and $\pm 1$ per cent at the higher concentrations, there being a steady graduation between the two levels of precision. The measurement is not affected by temperature, pressure, particle size or most common variations in chemical composition. As indicated, it is dependent on the density of the material, and compensation for this is normally essential. It has been used on belts, in bunkers and hoppers, and in solids-conveying pipelines.

The main reasons for using the techniques are:

(a) it measures hydrogen (and hence water) concentration in a larger volume
(b) its performance is largely independent of the properties of the material being analysed (boron and lithium interfere)
(c) its sensing head can be of very rugged construction.

Reasons against using it are:

(a) it employs fairly active radio isotope sources
(b) it is specific only to hydrogen not water—hence hydrogen containing solids or liquids interfere

(c) the sensing head is bulky and heavy, due to the shielding required against the radiation emitted
(d) since it induces radioactivity in the irradiated sample (albeit at minute levels) it is not acceptable for foodstuff examination.

Current applications include the determination of water in coke (directly in massive bunkers), glass sand, iron ore sinter concrete and wood chips—the last being an example of a successful application in the presence of a hydrogen containing solid.

### 5.2.9 Equilibrium Relative Humidity

A substance capable of taking up or releasing water vapour to the gas surrounding it—normally air—will do so until the vapour pressure of moisture remaining or contained in that substance equals the vapour pressure in the gas. Thus, if such a substance is allowed to come to equilibrium with an isolated volume of gas, the moisture content of the gas—determined by a suitable sensor from the range described in Chapter 2—can be related to the moisture content of the material. The relationship between the equilibrium relative humidity and the water content of a substance can be plotted as a curve called an adsorption isotherm. As the name implies, the relationship depends on temperature, and a series of such curves are normally plotted, covering the expected temperature range. These isotherms are highly dependent on the nature of the material and have to be determined experimentally. The term 'water activity' is commonly used on equilibrium humidity measurements. It is simply the equilibrium relative humidity expressed as a fraction rather than a percentage—65 per cent equilibrium relative humidity equals a water activity of 0.65. To translate this into practice, this corresponds, for example, to a moisture content at ambient temperatures of about 30 per cent for jam but only 8 per cent for an oil seed such as rape.

The time required to reach equilibrium relative humidity at ambient temperatures may be of the order of several minutes or even longer. At higher temperatures the exchange of moisture between gas and solid and within the solid is more rapid. However, it is obvious from the above description of the technique that in its most precise form it is difficult to apply on-line. A representative sample must be taken, placed in an enclosure full of gas—which may need predrying at low moisture concentrations—and allowed to equilibrate at a known temperature before a relative humidity can be measured. The author knows of no actual on-line applications using this method—although it could be resorted to if it were necessary to determine very low moisture concentrations in solids on-line ($<0.01$ per cent), where no other method is applicable. However, if the moisture content of gases passed through driers, bled out of conveyor lines or drawn from storage hoppers is measured, practice has shown that this can be reliably related to the moisture content of the solid material. The gas need not in all cases be drawn from the material; a probe immersed, say, in stored grain can with a small sinter filter create its own volume of equilibrium or near equilibrium gas and be used to determine the moisture content of the solids surrounding it—just as the aluminium oxide probe can be used to determine the moisture content of liquids by immersing them (see Section 3.2.2).

Concentrations of water as low as 0.01 per cent by weight in the solid material have been determined on-line by the use of the latter method. Where other condensable volatile compounds are likely to be given off in a drying process, it is essential to use a water specific sensor in the gas stream, such as the silicon or aluminium oxide varieties. The accuracy achieved depends on how well the adsorption isotherms are known for the particular substance, an accurate temperature measurement of the material or gas temperature, and how nearly equilibrium conditions are achieved. The last point must be qualified by saying 'or if not at equilibrium, how reproducible is the shift from equilibrium?'.

The advantages of the technique are:

(a) it is as specific to moisture as are the gas sensors used
(b) it can cover a wide range of moisture in solids, and may be the only applicable technique at very low moisture concentrations
(c) it is simple to apply on-line in its 'dynamic equilibrium' form
(d) it is inexpensive.

The disadvantages are:

(a) the adsorption isotherms have first to be independently established, and these may be sensitive to composition changes in the material being analysed
(b) the temperatures of the solid or gas need to be determined precisely
(c) equilibrium must be reached or approached to a reproducible degree if the measurement is to be valid.

The technique is widely used for dryer control of a broad range of materials such as foodstuffs, polymer chips, pharmaceuticals, textiles and fine chemicals. It is also used for monitoring the condition of materials such as grain stored in piles or hoppers, where the sensor is placed in a fine sinter 'cage', allowing the gas moisture content to be determined by pushing this about 1 m into the pile or hopper.

### 5.2.10 Other Methods for Determining Moisture in Solids

There are a number of methods which could in theory be used to determine moisture in solids on-line, although no application is known to the author. The colour of fine powders is often highly dependent on water content—almost entirely due to reduction of multiple scattering —and hence diffuse reflectance in the visible region of the spectrum could be applicable to certain problems—standard on-line diffuse reflectance sensors providing the hardware. Time domain reflectometry (TDR)[11] is a method of determining the dielectric properties of powdered materials at frequencies between 1MHz and 1GHz—thus covering the regions described in Sections 5.2.4 and 5.2.5. It measures the propagation velocity of pulses along coaxial and parallel transmission lines. It is apparently density -independent over about a 10 per cent change in bulk density, and has been mainly applied to soil analysis. No on-line application is known, although the technique when using parallel transmission lines could be applicable to the on-line situation. The noise emitted from powders being mixed or conveyed can be highly dependent on moisture content—both in sonic and ultrasonic frequency regions. These can be simply monitored using industrialised acoustic equipment, although the interpretation of the signal may be difficult. Electrical conductivity has been widely used for the off-line determination of moisture in solids—particularly in the construction and building industries. The problem with its use on-line is that of establishing good contact between the electrodes and the material—this could be overcome in suitable applications with an electrodeless system, similar to that used with liquids. Fully automatic chemical Karl–Fischer moisture systems are available for laboratory use, and these could be incorporated with an automated sampling system to give on-line sensing, although no strictly on-line applications have been located, probably because alternative solutions are available and easier to apply and use in a process plant environment.

## REFERENCES

1. Cornish, D. C., Jepson, G. and Smurthwaite, M. J., *Sampling Systems for Process Analysers*, Butterworths, London, 1981, pp. 96–258..
2. Merks, J. W., *Sampling and Weighing of Bulk Solids*, Trans Tech Publications, Clausthal-Zellerfield, FRG, 1985, pp. 127–230.

3. Shinskey, F. G., *Instrumentation Technology*, **15**(9) (1968) 47–51
4. Kraszewski, A., *J. Microwave Power*, **15**(4) (1980) 209–20.
5. Kent, M., *J. Microwave Power*, **7** (1973) 189–94.
6. Meyer, W. and Schilz, W., *J. Phys., D. Appl. Phys.*, **13** (1980) 1823–30.
7. Truxall, F. W., *Proc. Inst. Soc. Am.*, **19**(3) (1964) Paper 24.1.64, 12pp.
8. Willis, H. A., *Advances in Infrared and Raman Spectroscopy—2*, Heyden, London, 1976.
9. Richmond, I. J. and Mulward, R. C., *European Spectroscopy News*, **30** (1980) 49–50.
10. Van Bavel, C. H. M., *Humidity and Moisture*, *4*, Reinhold, New York, 1965, pp. 171–83.
11. Dalton, F. F., Herkelrath, W. N. Rawlins, D. S. and Rhoades, J. D., *Science*, **224** (1984) 989–90.

# CHAPTER 6

# *On-line Determination of Moisture Form*

Water can exist in solids, pastes, slurries and emulsions in differing forms. The simplest case is where water is present in an 'immiscible' liquid such as fuel oil. Part will be present as a true solution, the rest as 'free water' droplets (albeit saturated with the liquid) in differing degrees of dispersion. When solids are introduced, then the variety of possible forms increases—water of crystallisation, adsorbed water, water trapped in miscelles and various polymeric forms, to name a few. The basic variables of water form are:

(i) degree of hydrogen bonding—varying from near 100 per cent in pure water just above 0° to very low values with structurally isolated water molecules. The degree of bonding depends on the chemical environment and upon the temperature—although it is predicted that only 20 per cent or bonds are broken at 100°C
(ii) short-term order—how many water molecules the selected molecule is regularly associated with and in what geometrical arrangement
(iii) long-term order—how closely the structure of the water approximates to the normal close packed models of water or ice.

In many industries—particularly foodstuffs and for the future biotechnology—it is important to know not only how much moisture is present, but in what form it exists. In the on-line sensing field this is a relatively new requirement, so this short chapter is largely concerned with discussing potential rather than already applied techniques. This chapter will follow the form of the rest of book and examine this determination according to the phase of the sample.

## 6.1 GASES

Water can exist in a gas as vapour, droplets or ice crystals. To determine the droplet or crystal concentration can be difficult. If they are finely and uniformly dispersed, then light scattering techniques are appropriate. While conventional on-line turbidimeters or nephelometers can be used, some form of air flushing of the windows may be desirable to avoid deposition. With the simple adaptation of multiple angle scattering devices used for determining water in oils[1], discrimination against the effects of dust is possible. At higher concentrations the gas stream may be passed through a cyclone—warmed if necessary—and the flow rate from it measured. If possible, it is simplest to raise the temperature of the gas so that all the water is in the vapour phase, and if the concentration of both phases are required, carry out a gaseous moisture determination before and after heating. This determination is rarely required—more normally the presence of water droplets or ice crystals is a source of error in the determination of total water in the gas stream.

## 6.2 LIQUIDS

Here the water can exist as dissolved water, droplets or ice crystals. The same problem exists with liquids as with gases—if the water is in large, non-uniformly dispersed droplets, then uniform dispersion followed by determination by light scattering is the best solution. Infra-red ATR analysis (see Section 3.2.1) has proved effective in determining water in dispersed droplet form in oils and paraffins. Dissolved water can then be determined with an aluminium oxide probe—although in the presence of finely dispersed water the dissolved water content will be the saturation value at the measuring temperature.

## 6.3 SLURRIES, PASTES, SUSPENSIONS AND SOLIDS

It is with these materials that the full complexity of the problem is present, and where the industrialist is often most interested in the determination. The techniques which may be of value are now examined.

### 6.3.1 Infra-red Reflectance

The O–H absorption bands of liquid water are broad due to hydrogen bonding. When the water is present in different forms, the degree of hydrogen bonding changes, and the shape of the broad absorption bands shows corresponding shifts. If the degree of hydrogen bonding is low—for example in an isolated water of crystallisation molecule—the bands may sharpen considerably, and the phenomenon of a sharp peak superimposed on a broad band can be observed. To use infra-red diffuse reflectance, the change in the shape of the bands with changing water form has to be observed; this would involve laboratory investigation, using a scanning or interferometric spectrometer with diffuse reflectance attachment, which allows the whole absorption band profile to be recorded. When the necessary wavelength correlations with water form have been established, narrow band interference filters can be used in the on-line diffuse reflectance analyser to duplicate the measurement. Normally an instrument with three or more filter positions would be required, and as measurements may be required on the sloping wings of the absorption band it is essential to ensure that the filter temperature is closely controlled to avoid minor wavelength shifts. Giomini[2] has in a recent paper described the use of infra-red absorption for characterising water inside reverse miscelles, and the study is typical of those required to provide the information necessary for using diffuse infra-red reflectance for this determination on line. The absorptions associated with the weak hydrogen bonds are present in the far infra-red region, which is not easily accessible for on-line analysis, although laboratory measurements in this region may be of value for calibration purposes.

### 6.3.2 Capacitance Sensors

The dielectric constant of water changes with its form—and the change from one form to another varies with the frequency of measurement. Thus the measurement of the capacitance of a material containing water at differing frequencies can give an indication of the relative amounts of different forms of moisture present. The dielectric constant, when plotted against frequency, shows up to three plateau regions—roughly corresponding to audio, radio and microwave frequencies—with transition zones of varying sharpness in between. The microwave plateau is generally ascribed to 'free water', the radio frequency and audio regions to bound water in varying forms. Considerable work has been carried out measuring the dielectric properties of biological materials, and such information[3,4] may be of value in interpreting the data obtained. The measurement of the capacity of the materials at differing frequencies normally presents no more problem than with a commercial sensor at a fixed frequency—except that suitable frequency sources and bridges have to be used. In the microwave frequency (1–10 GHz) the technique merges with microwave attentuation—see Section 6.3.4. As with infra-red sensing, a laboratory investigation is required to establish the relationships—then the determination on-line becomes relatively simple, with the capacitance key measured either simultaneously or sequentially at two or more frequencies.

### 6.3.3 Low Resolution Nuclear Magnetic Resonance Absorption (NMR)

High resolution NMR can give considerable information on the form of water, as the proton absorption frequencies are highly dependent on the chemical environment of the protons. However, it is not currently practicable to use this on-line, and the frequency shifts are not sufficient to be detectable with the low resolution on-line instruments. The relaxation time is measurable with the low resolution technique, and this does depend on the chemical and physical environment of the proton[5] and hence may yield information on the form of the water.

Relaxation times have been studied for water and ice over varying temperatures, and the results used to understand the dynamic properties of these materials. As with the other techniques described in this chapter, their use to determine moisture form on-line is highly dependent on laboratory calibration work with the materials concerned. However, certain empirical correlations are possible; for example, the higher the viscosity of the environment of the water molecule, the longer the relaxation time; a similar correlation exists with degree of hydrogen bonding, although in water associated with solid structures, the sheer geometrical limitations may be the dominant effect. The related technique, electron spin resonance (ESR) has also been used to investigate the state of water in heterogeneous systems[6].

### 6.3.4 Microwave Attenuation

Microwave attenuation in materials is effectively dependent on the dielectric constants of the materials at the differing frequencies, but because of the differing hardware used for measurement, is treated in a separate section. The factor which varies is the speed with which polarisation of the water molecule occurs in the electromagnetic field. Thus, if attenuation is studied as a function of frequency, information can be obtained as to the form of water present in a material.

In pure water the relaxation frequency is around 17 GHz, but as the degree of binding increases, the frequency drops—in tobacco, for example, the bound water resonating at about 400 MHz[7]. Thus, by measuring the attenuation at one or more selected frequencies, the concentration of moisture in differing forms can be determined. However, here again the process control worker is forced to depend on the laboratory results obtained with the materials under examination, since the correlation of attenuation with water form has a complex theoretical basis which is not readily applied to even more complex natural materials. Additional complications are the difficulty, except in specialised laboratories, of measuring microwave attenuation spectra, and the non-availability of on-line microwave sensors at other than standard frequencies.

### 6.3.5 Thermal Techniques

Measurement of weight loss or the water vapour concentration in drying gas as a function of temperature can be used to determine the differing forms of water present in materials. Weight loss itself has the limitation of being sensitive to loss of other volatiles, but if the water vapour sensor used is adequately specific, such as the aluminium oxide variety, then water loss as a function of temperature can be readily measured. It would be extremely difficult to achieve such measurements on stream, but if only the relative amount of water in differing forms is required, a volumetric sample can be taken and placed in a suitable cell which is, together with its flushing gas, heated in a suitably programmed manner. A microprocessor would be required to sequence sampling, cleaning and temperatures, with a suitable read-out giving either relative weight loss at selected temperatures, or if it is known, the relative concentration of water in the different forms. No commercially available on-line systems are known, but, if the need were real, then the on-line interfacing should not be an insuperable obstacle for the majority of materials.

## REFERENCES

1. Pitt, G. D., *Electrical Communication*, **57**(2) (1982) 102–6.
2. Boicelli, C. A., Giomini, M. and Giuliani, A. M., *Applied Spectroscopy*, **38**(4) (1984) 537–9.
3. Grant, E. H., *NATO Advanced Study, Series A.*, **A49** (1983) 170–94.
4. Takashima, S., Gabriel, C., Sheppard, R. J. and Grant, E. H., *J. Biophysical Society*, **46** (1984) 29–34.
5. Klose, G. and Steizner, F., *Biochimica et Biophysica Acta,* **363** (1974) 1–8.
6. Ebert, B., Elmgren, H. and Hanke, T.,*Studia Biophysica*, **91**(1) (1982) 19–22.
7. Kraszewski, A. J., *Microwave Power*, **15**(4) (1980) 209–20.

# CHAPTER 7

# *Electronics for Data Transmission, Conversion and Verification*

In the process control environment, where the majority of on-line moisture determinations have to be carried out, making the physical measurement is only the first (but vital!) step in achieving the desired control. The measured parameter has either to be converted to equivalent water content by analogue or digital means directly associated with the sensor head, or transmitted back to an analogous unit or computer at a more or less remote site, without significant corruption of the signal.

The use of dedicated microprocessors in conjunction with on-line water sensing is becoming widespread for very good (rather than fashionable!) reasons. Firstly in using many techniques—such as capacitance or microwave attenuation—other parameters such as temperature or bulk density have to be measured and the values inserted into a predetermined equation relating the two or three quantities to water concentration.

This is most effectively carried out by a microprocessor since it has the additional advantage that the constants in the equation can readily be updated, if this is required, by subsequent calibration checks or a change in the material being sensed. A current trend is to supply a control output in the dedicated unit—since many moisture measurements are used to control simple unit processes such as dryers, where an integral unit has much to recommend it.

## 7.1 DATA TRANSMISSION

The output of many sensors used in the measurement of water is in analogue form—although with many this is more a tradition than a necessity; for example sensors measuring capacitance, such as the aluminium oxide or silicon, can as readily be designed to give a frequency output as an analogue one. With an analogue signal, this may be converted if necessary to a low impedance source, and transmitted in traditional format as a 4–20 mA signal. However, such signals are more prone to electromagnetic interference than either frequency or especially digital signals and there are sound reasons for commending the alternatives. The availability of low-powered, potentially intrinsically safe analogue to digital or voltage to frequency converter chips allows the signal to be converted at the sensor itself. At least one of the commercially available aluminium sensors (Endress & Hauser) uses a pulse-modulated form of transmission.

Where a large number of analogue signals are required—for example temperatures, or multiple moisture monitoring points in a grain store—then the use of a multiplexed analogue to digital converter becomes the most economic, as well as the most reliable, solution—cable costs can be up to £25 per installed metre!

The construction, routing and protection of data transmission cables is a major subject on its own, and readers are referred to standard works on the subject.[1] However, given the correct electrical characteristics of the cable and its terminations, the routing of cables can be largely based on common sense—which seems so often to be neglected. Running them alongside heavy motor starters, across fork lift truck routes or over heated ducts, seems obviously to be bad practice, but industrial experience shows many such errors being perpetrated, often for historical rather than planned reasons.

In the use of sensors in hazardous zones, the transmission cables have to conform to the required codes of practice—and again readers are referred to standard works on this complex subject.[2] Zener barriers have, at least in the UK, been the most widely used form of protection, but the use of galvanic isolators is increasing, and they are quite readily used with pulse or frequency signals. The use of fibre optic transmission is obviously very attractive in such zones, but currently is largely restricted to digital data transfer in distributed systems. There are various fibre optic based moisture sensors under development[3]—and when these become accepted, the optic itself is the obvious means of data transmission, as it would be with primary networks for sensors and actuators. However, fibre optics are at their most effective transmitting digital rather than analogue information, and so the need to carry out a sensor analogue to digital conversion may still exist. The use of telemetry for data transmission, especially on large sites such as tank farms or remote monitoring stations, is also an established technique. Here again digital transmission is used, with very low error rates due to inbuilt data verification (typically between 1 in $10^6$–$10^9$). It is, however, used less in the strict process control context, possibly due to higher levels of interference and a smaller differential in cost saving. However, as low-cost chips become available with the capability of carrying out the necessary data verification routines, the technique is likely to become more important in this area.

## 7.2 DATA CONVERSION

In the simplest analogue system the output from the sensor is suitably scaled, then used as the input to an on/off proportional or PID controller, which in dryers, for example, can control heat input, gas throughput or solids residence time. The next stage in complexity is to use analogue linearisation, with scaling and adjustable zero offset as the input to the controller. At this stage it begins to make sense to consider the possibility of using a microprocessor based single loop controller, provided it has the capability to carry out these operations on the incoming signal.

The next stage in complexity is to use a microprocessor based controller, which is dedicated to a particular type of sensor or a limited sensor range. Typically the outputs of the moisture sensor and a temperature sensor—either resistance or thermocouple, depending on the demands of the application—are used as the inputs to the controller, which carries out the linearisation (either through look-up tables, or preferably a polynomial or analogous equation), scaling and offset and also corrects for the effect of the temperature of the material being sensed on the moisture reading. As has been previously mentioned, there is a move towards providing simple single loop-control with such a unit, and it may have the capability of being remotely controlled in a distributed system, so that, for instance set points and control constants can be changed from a central control unit.

The final degree of complexity is to feed the input from the moisture sensor, together with the input for temperature, flow and even material type, into a local intelligent node in a distributed control system or the process control computer, which can control a number of parameters and achieve optimum control of the plant.

## 7.3 VERIFICATION

In all the systems there must be the capability of locally updating the calibration data of the sensor when this is checked. This may range from potentiometers on the simplest system described above, through an electronic checkout of the whole system in dried gas to a two point calibration check involving standards, flushing and sequencing, plus automatic updating of the constants of the calibration equation—tasks ideally suited to a local microprocessor with simple sequencing and control facilities. The type of dew point sensor (Section 2.3.3) that operates very close to the dew point all the time, is normally periodically heated up to remove volatiles and this again is a function well suited to a local microprocessor. Similarly, the vibrating crystal sensor (Section 2.3.10) requires regular cycling, and is ideally suited to mating with a local unit. Such microprocessor based systems of data verification are now available from a number of commercial suppliers. Where this is not so, there are available a number of programmable sequence controllers with simple analogue input/output and the required mathematical capacity to carry out data conversion and verification. If sufficiently clean working conditions are available, a personal computer with analogue input/output facilities can be used, and a number of software packages containing the required facilities are now becoming available.

## REFERENCES

1. American Petroleum Institute Standard API RP550, 1974.
2 Bass, H. G., *Intrinsic Safety*, Quartermane House Ltd, Sunbury, UK, 1984.
3 Russell, A. P. and Fletcher, K. S., *Analytica Chimica Acta*, **170** (1985) 209–16.

# CHAPTER 8

# *Calibration of On-line Moisture Sensors*

Calibration for on-line moisture sensors can be conveniently divided into three areas. The first is the initial characterisation of the quantitative performance of the sensor type over a wide range of concentration, pressure and temperature. The second is on the production line, where the calibration of individual sensors is checked before despatch to the customer. The third is the regular checking on calibration required in plant use to ensure that the sensor is supplying the control system with adequately meaningful data.

## 8.1 GASES

The readily obvious (but not so readily obtained) initial requirement for sensor characterisation is the availability of accurate standards of known water content, covering the full operating ranges of concentration, pressure and temperature that the sensor can reliably cope with. With gases, considerable research and development effort has been undertaken by bodies such as the UK National Physical Laboratory[1] and the US National Bureau of Standards to establish primary moisture standards using gravimetric techniques. To transfer this information to other organisations, very precise mirror dew point hygrometers[2] have been developed as secondary standards. However, as has already been stressed, precision, selectivity and both short and long-term reproducibility are of primary importance in process control sensing—high accuracy being a secondary factor in the majority of applications—one exception being where mass balances are being attempted in plant modelling. The technique adopted by many manufacturers (and their production and maintenance departments) is to make use of the available published information[3] on the saturated vapour pressure of pure water at controlled temperatures. The calibration gas is first saturated with moisture at the selected temperature—for example, by bubbling through water columns at precisely controlled temperatures. The gas then passes through an aerosol separator and is then raised to a higher temperature to avoid the possibility of condensation. This moist gas stream is then diluted with a highly dried gas at carefully controlled flow rates. Where low moisture contents are required there will be several stages of dilution. Great care has to be taken not only to dry the gas completely down to a level that will not affect the lowest concentration standards (molecular sieves and cryogenic cooling are now being used in preference to phosphorus pentoxide—although the latter may be used as a final polisher of the dried gas) but also to ensure that the materials of construction of every gas-contacting component in the system do not adsorb or desorb water vapour in sufficient quantities to invalidate the dilution procedures. Section (2.2) on sampling systems for gaseous moisture sensors outlines the problems involved and the solutions available. A mirror type dew point or frost point hygrometer is widely employed in the final mixed gas line to confirm that the dilution has been successfully achieved and to sense for the initiation of leaks or other extraneous moisture sources within the system. One alternative to the saturated vapour pressure/dilution standardisation technique is the use of permeation tubes. The tubes, made of PTFE or other well-characterised and suitable plastic, are immersed in a temperature-controlled waterbath. They allow water vapour to diffuse through their walls at well-established rates. Subsequent dilution with dried gas is used to produce lower concentrations, as with the saturated vapour pressure method. While the diffusion tube technique is very convenient, it is not directly traceable back to internationally accepted tables, as is the vapour pressure technique. However, this technique can produce moisture concentrations directly at values as low as a few ppm. Another alternative is to use diffusion along the bore of a polished stainless steel tube. This is not subject to the variabilities inherent in the use of plastic materials, and is capable of high and long-term stability—the manufacturer of such a system offering a 25-year guarantee[4]. For production checking, a simplified system, often microprocessor-controlled, is normally all that is required to run each sensor through a regular test series. 'On the plant' checking requires not only a portable calibrator, which may need to be intrinsically safe, but also the design and skill to ensure adequate coupling and purging. Such are the magnitude of these problems that removal and checking of the sensor in a plant laboratory, or building in a calibrator in critical applications, are generally the preferred choices.

Calibration and characterisation, once a supply of gas streams having known moisture concentrations has been ensured, involves a number of other important factors. The sensor response must be measured against rising as well as falling concentrations, to check for hysteresis, and the response times to large upward and downward step changes are used to investigate response times and memory effects. The effects of pressure and temperature must also be investigated in detail—particularly the acceleration of any longer-term calibration drifts at higher operating temperatures. Selectivity was described as one of the most desirable characteristics for any process control sensor, and this too must be checked at the calibration stage, using the most commonly found potentially interfering gases. Particular customer requests will have to be met at the production stage, thus adding to the applications knowledge of the sensor manufacturer. Longer-term stability can be checked by accelerated ageing at the development stage and later by field performance. Such stability should be indicated on the information supplied with the sensor—typically as per cent full scale drift per year under specified concentration and operating conditions. Once a sensor type has been well characterised, production calibration can be simplified, and may be simply a two- or three-point calibration check providing an accompanying certificate on performance. Similar checking is usually adequate for 'on plant' performance confirmation—indeed this is often reduced to response to 'dry' and 'wet' air, the latter being most readily achieved by placing the sensor in a closed space in equilibrium with a compound having known and stable equilibrium relative humidity, such as a number of salts (see Table 1). This technique is not suited to lower concentrations, where a simple calibrator using a diffusion tube or membrane is more appropriate and accurate.

Two areas presenting special problems on the calibration of gaseous moisture sensors are the determination of very low (1 ppm) and very high moisture contents. At very low moisture contents, the main complications are drying the gas sufficiently to act as the base material for water addition, and ensuring that the standards are not contaminated by desorbed water. Both these problem areas have been discussed previously in this chapter and in Section 2.2 in the chapter on gaseous moisture sensors. Producing high water content gas standards is largely a matter of good laboratory practice—gravimetric or volumetric additions into volumes at reduced pressure and raised temperature being generally

**TABLE 1**
Relative humidity standards

| *Salt* | *Percentage relative humidity above saturated solution at 25°C* |
|---|---|
| Potassium sulphate | 97·0 |
| Ammonium dihydrogen phosphate | 92·7 |
| Potassium chromate | 86·5 |
| Sodium chloride | 75·1 |
| Sodium dichromate dihydrate | 53·7 |
| Magnesium chloride hexahydrate | 32·7 |
| Lithium chloride monohydrate | 11·3 |

applicable. The main problem is ensuring that saturation does not occur and water does not condense out from the gas—local cool spots being particularly troublesome. It is more difficult to produce accurate high moisture concentrations in continuously flowing gases but blending of steam and base gas flows is possible, again with adequate steps to ensure no localised cooling occurs, and that any pressure reduction does not cause overall cooling below the dew point. Monitoring of both temperature and pressure is normally essential both to ensure standard integrity and to replicate the actual process conditions, which often dictate measurement at high temperatures and non-atmospheric pressure.

## 8.2 LIQUIDS

Here again the first step is to ensure an adequate set of standards. The obvious way of producing these is by dissolving the required quantities of water in the base liquid, having first ensured that the water content of the base liquid is low enough not to interfere. Subsequent volumetric dilution can be used to produce lower standards, using normal good laboratory practice to avoid moisture contamination or loss by evaporation of the water or base liquid. Volumes of water as low as $10^{-2}$ ml can be accurately dispensed—giving in a litre sample a concentration of 10 ppm by volume directly. Another source of water loss can be adsorption on the containing vessel walls—but with liquids this is only likely to occur at very low concentrations. Where flowing standards are required, recirculation in a sealed loop (with a facility such as a bellows or diaphragm to take up any volume changes caused by temperature shifts) is feasible, using stainless steel as the construction material. Careful cleaning and drying of the loop before use and between standards is necessary. Water can be added by injection through a small elastomeric seal (similar to those used in gas chromatography) which can be adequately isolated from the liquid by an air gap or by the use of normal metering pumps, made of suitably inert materials.

Two problem areas exist in the production of liquid standards. The first is ensuring that the base liquid used has a low enough water content so as not to interfere with the calibration. Drying agents for liquids are highly dependent on the chemical and physical properties of the base liquid—for example, phosphorus pentoxide and finely dispersed sodium metal have both been used to dry liquids to very low moisture contents, but both can react violently with many liquids. Molecular sieves can also be used to scavenge low concentrations of water from many (particularly organic) liquids. One of the most important requirements when drying liquids is having the ability to follow the base liquid drying process. Infra-red absorption, aluminium oxide probes and conductivity are particularly valuable techniques with which to do this.

The second problem area in the production of liquid/water standards is where the water content exceeds the solubility of water in the base liquid—either throughout or within the required calibration range. Volumetric or gravimetric additions of water again present no significant problems—but ensuring that the water droplets are in the same degree of dispersion as that which will be encountered in the process definitely does. It is best to carry out the on-plant measurement in a 'fast loop', the scale and components of which can be copied in a laboratory rig. Similar means of dispersion, such as a centrifugal pump, side injection or an ultrasonic atomising unit, can then be employed in both situations. Care has to be taken to ensure that all the water added is dispersed and does not lie in a thin layer at the bottom of pipes or other components. This is best achieved by injecting the water into (or immediately upstream of) the dispersing device. Since some sensors, such as infra-red ATR, give a very different response for dissolved or suspended water, and as the equilibrium between the two can be highly temperature-dependent, it is essential to carry out the calibration at a controlled temperature. In a recirculating loop the energy dissipated by the pump can rapidly raise the temperature of the liquid in the loop. It is normally essential to include a cooling system in the recirculating loop. A water cooler is generally quite acceptable, with a proportional temperature controller acting on the flow rate of the cooling water to give temperature stability.

## 8.3 SOLIDS, PASTES AND SLURRIES

With such materials, the type and preparation of the standards used for calibration will depend on the method of water determination being used. With some sensing methods, such as neutron moderation, neither the chemical environment of the water molecules nor anything except the grossed non-uniformity of mixing will affect the calibration of the sensor. With some, such as capacitance, the chemical environment of the water molecules affects calibration, while with surface-sensitive techniques such as infra-red diffuse reflectance, both the chemical environment and water distribution throughout the material can change the calibration data.

The simplest method of preparing water standards for solids is to oven dry a sample to constant weight and then add known quantities of water to portions of it. The water may be added gravimetrically or volumetrically and then the standard mixed in a container which is sufficiently sealed to ensure water is not lost to or gained from the atmosphere during the mixing process. Mixing should continue until the response from the sensor is no longer time-dependent. It is difficult to give a more precise instruction, for the mixing time can vary from under a minute to several hours, depending on the material and method being used. Even with such precautions it is not always possible to achieve a standard sample giving equivalent sensor response per unit of water concentration as the material examined on line. One reason is that it is difficult to ensure complete drying of the base material—many organic powders, such as starches, cannot be dried to less than a few per cent moisture content without partial decomposition. A second reason is the difficulty in ensuring that the moisture distribution in the grains of the solid is identical (as far as the physics of the sensor is concerned) with that in the grains of the process material, even if mixing to a constant sensor response can be achieved. The third reason is that it may be impossible by mixing to ensure that the water molecules added take up the same chemical environment as those previously or already in the material—the so-called 'water form' effect. Because of these problems it is often preferable to calibrate the sensor using samples taken from the process line which have their water content subsequently determined in the laboratory by standard techniques such as Karl–Fischer titration, loss in weight or gas evolution following reaction with a suitable reagent such as calcium carbide. Where the technique is capable of examining small standard samples, such as infra-red diffuse reflectance, a set of these can be obtained and stored in sealed containers—although even with this technique ageing can affect the calibration. Where larger samples are required for the particular sensor type, care must be taken to ensure a representative sub-sample for the laboratory analysis—not too difficult if the material is a fine non-cohesive powder, but much less easy with a paste or a coarse fibrous solid such as sugar beet pulp or tobacco. Here the well-established rules of good laboratory sampling must be adhered to.

With slurries, where the main liquid phase is water, larger standards have to be made available so that they can be continuously mixed by pumping round a recirculating loop. This is normally essential to avoid segregation and settling. Since the water contents normally range between 40 and 95 per cent, small amounts of water left in the dried base solids are a less important source of error. Standards are usually made by mixing known weights of oven dried solid with known weights of volumes of water. With a recirculating loop it is convenient to make successive additions of solid, up to the lowest water content it is required to determine. If the main slurrying liquid is not water then calibration can present greater problems. Volumes of water can be added to the recirculating slurry but, particularly at low concentrations, the uncertainty of the distribution of the water between the liquid and solid phases can affect the calibration. In this case as well, using representative, subsequently analysed samples is the best route for calibration. Here yet again there is a need to ensure representative sub-sampling from the recirculating slurry in the loop. While a wide range of slurry samplers of varying complexity exists, they are generally too large in scale for a small recirculating loop (typically employing 25 mm ID pipes) and the author has found that a simple 'T' tapping on or close to the pump chamber (usually a centrifugal pump is the best choice with slurries) or in another area of very high turbulence can be relied upon in most cases to give representative sampling. On the scale of the pipework used in recirculating loops, complete diversion of a portion of the flow into a can or bucket is an effective—though sometimes messy—alternative.

For thick pastes the main problems are usually those concerned with the handling of materials. Providing the samples can be pumped, water additions can be made and recirculation used to achieve adequate mixing. Monopumps (or their near relatives) are effective means of moving many thick pastes in a recirculating loop. Sampling for subsequent laboratory analysis is usually only possible by complete diversion of the flow into a bucket, but recirculation must have continued sufficiently long to ensure that the whole loop is homogeneous—subsequent sampling from the same loop and material can check on this.

In all cases where off-line calibration is carried out for solid, paste or slurry sensors, it is normal to check the performance when operating on the process by taking representative samples from the process stream for subsequent laboratory analysis and comparison with the sensor output at the time of sampling. This is commonly known as 'calibration trimming'. It is essential yet again to stress:

(i) that truly representative samples are taken. This requires care and the use of adequate sampling techniques. The reader is referred to standard works on this subject[5].

(ii) that an adequate number of samples is taken to show up any scatter in results due to random sampling and analytical errors.

(iii) that a wide enough range in water content is covered to clearly delineate the calibration graphs. The latter is often the most difficult to achieve, for in a well-controlled process it may be impossible, for production or engineering reasons, to vary the water content of the process stream at will. In such a case calibration 'trimming' may be little more than a one point calibration check. If reasonable agreement is achieved between the pre-established calibration graph all well and good. If not, the whole question of slope, linearity and origin of the calibration graph remains open. With some sensors it is possible to check the zero point independently of a sample and this helps to clarify such a situation.

Where a regular one point instrument stability check is required, it is possible with some sensors, such as those using diffuse infra-red reflectance or neutron moderation, to use a material other than the normal sample. For example, hydroxyl containing polymer discs can be used to check the stability of an infra-red gauge, and mixed polythene (hydrogen source) and graphite chips a neutron moderation gauge. This does not take into account calibration shifts due to other factors such as changing quality of a natural raw material or process operations, which may affect the overall calibration.

## REFERENCES

1. *Humidity Measurement Standards for the UK*, National Physical Laboratory, Teddington, UK, 1985.
2. Michell Instruments, Leaflet 4000/3/84, Cambridge, UK, 1984.
3. *Handbook of Chemistry & Physics*, CRC Press, Boca Raton, Florida, USA, 1984, pp. D192–4.
4. Moisture Control & Measurement Ltd, Si-Grometer Brochure, Wetherby, Yorks, UK, 1985.
5. Merks, J. W., *Sampling and Weighing of Bulk Solids*, Trans Tech Publications, Clausthal-Zellerfield, FRG, 1985, pp. 127–230.

# CHAPTER 9

# *Selection of On-line Moisture Sensors*

A process engineer requiring an on-line water sensor is often faced with a bewildering variety of sensor types and manufacturers, all having their competing and sometimes contradicting claims. There is a natural—and often justifiable—tendency to seek out and use a sensor that has been applied, at least with some measure of success, to a moisture measurement similar to the one in hand. However, a brief scan of applications claimed by manufacturers to be 'successful' shows considerable overlap amongst the competing varieties and their manufacturers. Furthermore, a successful application may have involved an enthusiast beavering away for months or years getting the conditions just right before an adequate performance could be achieved, while a rival but untested (in the particular application) sensor could almost be plugged in, switched on and be ready to go. Similarly, a sensor may be successful when supported by maintenance and calibration effort which is just not available on another plant, while an easily calibrated sensor with low maintenance requirements already exists, but is as yet untested for the required application. Theoretical objections and selection on this basis alone can also present problems. For example, infra-red diffuse reflectance is only sensitive to surface moisture concentrations and hence, from a theoretical basis, should only be used when the surface is truly representative of the bulk of the material. However, it has been successfully applied in many cases where this is manifestly not true. In a large number of such cases a consistent correlation exists between the surface moisture concentration and that in the bulk of the material. While this may be considered undesirable since the correlation may shift if process modifications are introduced up stream, the ease of application and installation of the infra-red sensor has made it a preferred choice in such applications. With measurement to determine the end of drying, the surface sensitivity may sometimes be a positive advantage, since if the surface dries last a sharp end point is indicated by the infra-red gauge.

Sensor selection will be reviewed under the types of material analysed. Summaries are given in the selection Tables 2–4.

## 9.1 GASES (See Table 2)

This could be simplistically stated as: 'try silicon first, then aluminium oxide sensors second. If these do not work, start looking at the rest.' In practice this does form a good basis and the 'pros' and 'cons' under Sections 2.3.7 and 2.3.6 on silicon and aluminium oxide sensors form the starting ground. They are the most popular on-line gaseous moisture sensors, and for very good reason.

The first objection to the use of silicon and aluminium oxide sensors is that the aluminium oxide (and to a much significantly lesser extent silicon) sensors show slow drifts which seem to be associated with the basic physical structure of the sensor[1,2]. Various additives and methods of oxide preparation have been tried with the aluminium oxide sensors to reduce this effect and while they may be less in one manufacturer's probe than another, they mean that the facility must be available for regular performance checks, the frequency of which is dependent on the manufacture, the accuracy required and the operating conditions of the probe. Some gases, high temperature and high humidities tend to accelerate drift and hence demand more frequent checks. Under the latter conditions the preferred choice may shift to the high temperature/humidity thermo-electrically cooled dew point probe using capacitive detection of dew point formation (such as made by Endress & Hauser), cross duct infra-red absorption (other gases can be determined simultaneously) or some of the polymer based sensors which have a successful application record under such conditions (such as those made by Vaisala).

A related objection to aluminium oxide and silicon sensors is that at high temperatures and humidities the sensors need to be operated at a controlled temperature. In fact some degree of temperature control (or measurement plus correction) is desirable for most sensors under these conditions, if for no other reason than to ensure the sensor temperature does not approach the dew point of the sample stream. With infra-red gauges the calibration is a function of gas temperature, and this must in any case be measured and a correction applied unless the sensor is in a sample line giving controlled gas temperatures.

**TABLE 2**
Selection of sensors for gases

| *Required characteristic* | *Mechanical* | *Wet and dry bulb* | *Dew point* | *Electrolytic* | *Lithium chloride* | *Silicon* | *Aluminium oxide* | *Polymer* | *Crystal oscillator* | *Mass spectrometry or gas chromatography* | *Notes* |
|---|---|---|---|---|---|---|---|---|---|---|---|
| Trace water (100 ppm) | | | + | + | | + | + | | + | | |
| Low water (<10 per cent RH) | | ○ | + | + | + | + | + | + | + | + | |
| Medium water (10–90 per cent) | + | + | + | + | + | + | + | + | ○ | + | |
| High water (>90 per cent) | | ○ | + | | | + | + | ○ | | | |
| Fast response | | | ○ | + | + | + | + | | | | slower at low concentrations |
| High accuracy | | | | | | + | | | + | ○ | |
| High reliability | | | | | | + | + | | | | |
| High specificity | | | | + | + | + | + | | | + | |
| High stability | | | ○ | | | + | | | + | | |
| Low cost | + | | | | | | + | + | | | |
| In-stream | | | ○ | + | + | + | + | + | | | |
| Ruggedised | | | | | | + | | | + | | |
| Intrinsically safe | + | | | | | | + | | | | |
| Explosion-proof | | | | + | | | | | + | | |
| Ease of calibration | | | + | + | ○ | ○ | ○ | ○ | + | ○ | |
| Ease of application | | | ○ | | | + | ○ | | | | |
| Corrosion-resistant | | | ○ | ○ | | ○ | ○ | | | ○ | |
| Stands immersion | | | | | | + | + | ○ | | | |

+, Characteristic met by some models.
○, Characteristic may be met by some models.
Rough preference order (most preferred first): silicon; aluminium oxide; dew point; polymer; electrolytic; crystal oscillation; lithium chloride; mechanical; wet and dry bulb; mass spectrometry; gas chromatography.

The second main objection to silicon and aluminium oxide probes is that they are attacked by corrosive gases. Silicon is generally somewhat more inert than the oxide type probe and should have a wider range of application. Where corrosion or greatly accelerated drift due to chemical effects is a problem, infra-red absorption can be an attractive alternative, for infra-red transmitting window material can generally be found that can stand up to most corrosive gases, and gaseous flushing or window heating can also be used where helpful. The mirror or capacitive type dew point sensors can also be made using materials such as gold, platinum or ceramics which are resistant to many corrosive gases. Tantalum oxide based sensors, analogous to the aluminium oxide ones are also available commercially, and may present an alternative, since tantalum is much less prone to attack than aluminium. However, these have not been widely used, in part at least because they show a smaller change in capacitance with water content than the aluminium oxide varieties. Some 'one off' solutions to corrosive gas problems exist—for example, the phosphorus pentoxide sensor, while being rapidly attacked by moist chlorine gas, is a good choice for monitoring traces of moisture in well-dried chlorine.

The final objection to silicon and aluminium oxide sensors is that they are not absolute sensors—they need calibration and the calibration needs checking from time to time. Where the inbuilding of a calibration system or regular withdrawal for laboratory checks is not feasible, either of the so-called 'absolute' methods can be used—dew point sensing with cooled mirror or capacitive sensing of the dew formation or the phosphorus pentoxide sensor using the electrolytic decomposition of phosphoric acid formed by water absorption. However, both of these may be less 'absolute' than is at first apparent. Deposited salts or other hygroscopic materials can affect the calibration of the dew point type sensor (some manufacturers claim to minimise this effect by design and mode of operation) while the back electrolysis of the phosphoric acid back to phosphorus pentoxide is not quite as stoichiometric (especially as time of operation of the sensor increases) as the equations used with the manufacturers' literature might suggest. Nevertheless, both techniques offer the possibility of calibration free working, with the bias in favour of the dew point sensor where it is applicable.

There remain certain 'special' gas moisture sensors. The vibrating crystal sensor provides the 'Rolls-Royce' of gaseous moisture sensors, with built in calibration and checking facilities offering high reliability, high stability and low limits of detection at a high price—triplicated silicon detectors using a simple voting microprocessor verification system might be an economic alternative where high reliability is essential. Infra-red sensors are particularly attractive where cross-duct sensing is convenient or essential or where other compounds require determination. Because of its selectivity, ruggedness and speed of response, it is a method that is growing in popularity. Where water is required as just one of a number of compounds in a process stream, on-line mass spectrometry or gas chromatography may be used to determine moisture content, but because of the problems encountered in their use, such as the need for sampling systems, maintenance and such-like, they will rarely, if ever, be used for simple moisture determination.

## 9.2 LIQUIDS (See Table 3)

Liquid moisture sensor selection appears to be less clear cut than it is with gases. The first step is to ascertain if the water is present (i) only in solution through the expected working range of concentration, (ii) mainly as an immiscible phase, or (iii) as a mixture of the two, with saturation being reached within the expected working ranges of concentration and temperature.

Where the water is present in solution at low or very low concentration, the immersed aluminium oxide probe is a preferred sensor, providing its materials of construction are inert as far as the liquids involved are concerned. It is necessary to know the temperature of the liquid if significant variations in it occur. The probe essentially determines the equilibrium relative humidity of the liquid, and in some applications the technique can be used with the sensor in the enclosed gas phase above the liquid. However, unless this is strongly preferred due to the liquid attacking the sensor, problems with temperature differentials and reaching true equilibrium make the direct immersion of the probe the preferred choice. Where higher concentrations of water, sensor attack or difficulties in checking calibration rule out the use of aluminium oxide probes, infra-red absorption is an attractive alternative. It is capable of covering the concentration range from a few parts per million up to water as a major component with high selectivity and stability. The main limitations may be cost—it is somewhat more expensive than the simple probe—and interferences caused by absorption bands of other hydroxyl-containing liquids in the process stream. With the infra-red method the technique of using multi-attenuated total reflectance (ATR) is a very powerful one for

**TABLE 3**
Selection of sensors for liquids

| *Required characteristic* | *Infra-red* | *Aluminium oxide* | *Process refractometer* | *Physical property* | *Equilibrium relative humidity* | *Mass spectrometry or gas chromatography* | *Notes* |
|---|---|---|---|---|---|---|---|
| Low water (<0·2 per cent) | + | + | | | + | + | |
| Medium water (<10 per cent) | + | | + | + | + | + | |
| High water (>10 per cent) | ○ | | + | + | ○ | ○ | |
| Fast response | + | + | + | + | | ○ | |
| High accuracy | | | ○ | ○ | | ○ | |
| High reliability | | + | | | + | | |
| High specificity | ○ | + | | | ○ | ○ | ERH depends on sensor |
| High stability | ○ | | | + | | | |
| Temperature-insensitive | + | | | | | + | |
| Low cost | | ○ | | ○ | + | | |
| Non-contacting | | | | | + | | |
| In-stream | ○ | + | + | + | | | ATR infra-red only |
| Ruggedised | | | ○ | ○ | | | |
| Intrinsically safe | | + | | ○ | + | | |
| Explosion-proof | + | | + | | | | |
| Ease of calibration | + | + | | | | + | |
| Ease of application | ○ | ○ | ○ | ○ | | | |
| Corrosion-resistant | ○ | | + | ○ | | ○ | |
| High penetration | ○ | | | ○ | ○ | | |

+, Characteristic met by some models.
○, Characteristic may be met by some models.
Rough preference order (most preferred first): aluminium oxide; process refractometer; physical property; infra-red;

moisture determination, with the probe directly immersed in the process stream, reactor or fast loop.

Where higher concentrations of water are to be determined, measurement of a physical property of the solution may represent a simple and rugged method.

Density, dielectric constant, refractive index or conductivity are preferred physical properties, provided that the physical property of the base liquid remains sufficiently constant to permit their use. Commercially available sensors cover most physical properties, although dielectric constant appears poorly represented for liquids. Entrained gas will interfere with nearly all except refractive index, and temperature will need sensing if it varies significantly.

If the moisture is present in the liquid as an immiscible phase, the preferred method of determination at lower concentrations is to disperse the water uniformly and reproducibly into droplets by means of a centrifugal pump, high speed stirrer or ultrasonic emulsifier, and determine the droplet concentration by light scattering at one or more angles. At higher concentrations, capacitive or density measurements are more applicable and even with these it is desirable to disperse the drops before measurement. If a fast loop or other sampling system has to be employed, it should if possible be placed after the droplet dispersing device which will aid representative sampling as well as measurement. It has been demonstrated that an ATR infra-red probe is an effective way of selectively determining water droplets after uniform dispersion. The exact mechanism involved is not clear since ATR is a surface technique, but extended industrial experience confirms the validity of the application.

Where the water concentration reaches saturation in the concentration and temperature ranges expected, it will start at the lowest concentrations being completely dissolved in the liquid. This state will continue as the water concentration rises until the stability at the operating temperature is exceeded, when water droplets will form (assuming some form of supersaturation does not occur). If the process stream always passes through a means of uniformly dispersing the many droplets, a liquid ATR infra-red analyser can then determine the total concentration of water present—the calibration graph for water in diesel oil being typical.

Where the water concentrations being dealt with are higher, then density or capacitive techniques can be used after dispersion—but bear in mind that a break in slope of the calibration graphs is likely to occur at the saturation point. It may be possible in some applications to raise the temperature of a sample stream so that the saturation point lies above the highest expected concentration, thus simplifying the measurement. It should be noted that any method of measuring the moisture content by means of the equilibrium relative humidity of the liquid would not be a viable method of measurement when free water droplets are present.

## 9.3 SLURRIES, PASTES AND EMULSIONS

Because of the wide range of physical properties and water contents encompassed by slurries, pastes and emulsions, it is difficult to produce a clear and unambiguous guide to sensor selection. In many cases the overwhelming problem will be the material handling required to produce a suitable sample stream or even discrete samples for automatic on-line moisture determination.

With slurries the water content is generally high and sample handling problems are less evident. The most widely applied method for determining water in slurries is measurement of the slurry density. $\gamma$ Ray attenuation is the preferred technique in the more rugged industries such as mineral processing, although the vibrating tube gauge is to be preferred in many applications due to its greater precision and ease of installation. Care must be taken to ensure that there is no significant entrained gas in the process material (1 per cent by volume of gas causes approximately 1 per cent error in density which can equate to several per cent of water!). Where the solid density can vary widely, or is close to that of water (as in many biotechnology slurries), or entrained gas cannot be removed, other means of water determination must be sought. Dielectric constant (beware entrained gas) or conductivity may be useful alternatives and the viscosity of a slurry can be a sensitive indicator of water content if simple methods fail. NMR and neutron moderation with $\gamma$-ray attenuation or scattering for water-independent density variation correction are possible but expensive back-ups. Ultrasonic attenuation has been widely used to determine the solid content (and hence the water content) in sewage sludges, and it may have applications in other biologically based slurries—but yet again, beware of entrained gas.

For pastes, the sensors suggested in the following section on solids are generally applicable with a similar choice table. Since the moisture content is appreciable, segregation of water either at the upper surface (with solids more dense than water) or at the lower surface (in the reverse case) can occur and this would suggest a preference for the more highly penetratory techniques, such as capacitance, microwave attenuation or neutron moderation. However, on handling pastes it is extremely difficult to avoid entraining significant amounts of air or other gas—or to remove the gas once entrained. This would result in the need to employ density measurement and compensation—an undesirable complication. It has been found that infra-red diffuse reflectance, even though a surface technique, can give good results with moisture determination in pastes especially if some form of mixing is employed immediately upstream of the sensor. The rheological properties of many pastes are highly dependent on water content, and this may be a simple and attractive alternative where it can be applied, the power load on clay extruders being a cited sample.

Emulsions and suspensions have similar requirement to slurries—and indeed emulsions overlap with the liquid samples where immiscible water is present (Section 9.2). Some emulsions contain both dispersed immiscible liquid and solid (emulsion paints being a good example). Generally it is understood that with an emulsion or suspension the different phases are finally dispersed and that a degree of time stability is present. The emulsions may have small liquid–liquid density differentials, very marked non-Newtonian viscosity behaviour and variable solid to non-aqueous liquid concentration ratios. This makes density not the first choice of measurement method, requires viscosity measurement of water content to be applied with even greater care and can complicate the use of light scattering techniques. The smooth surfaces formed by emulsions often have high specular reflectance, making diffuse reflectance infra-red analysis difficult to apply and they tend to foul the optical surface of the infra-red ATR crystal (if they do not this is a good method for emulsions). One possible method is to use the specular reflecting surface and a mirror to allow multiple specular reflections in an analogous method to ATR, but the author knows of no on-line applications.

## 9.4 SOLIDS (See Table 4)

Here again a simplistic rule has some value. It is: 'Try infra-red diffuse reflectance—because of its ease of installation, application and versatility. If that cannot be used—try the rest'. Indications against infra-red diffuse reflectance are cost, non-uniform distribution of moisture in the bulk or grains of the solid, high infra-red absorbance over a wide spectral range and ease of formation of specularly reflecting surfaces. Very high water

**TABLE 4**
Selection of sensors for solids

| *Required characteristic* | *Loss in weight* | *Mechanical properties* | *Temperature difference* | *Microwave attenuation* | *Capacitance* | *Infra-red reflectance* | *NMR* | *Neutron moderation* | *Equilibrium relative humidity* | *Notes* |
|---|---|---|---|---|---|---|---|---|---|---|
| Low water (0·2 per cent) | | | | | | ○ | | | ○ | |
| Medium water (<10 per cent) | + | + | + | + | + | + | + | + | + | |
| High water (>10 per cent) | + | ○ | ○ | + | + | ○ | + | + | ○ | |
| Fast response | | ○ | ○ | + | + | + | + | ○ | | |
| High accuracy | ○ | | | | | ○ | ○ | | | |
| High reliability | | ○ | ○ | + | + | + | ○ | | | |
| High specificity | | | | | | + | + | + | ○ | ERH depends on sensor |
| High stability | + | | | | | ○ | + | | | |
| Temperature-insensitive | ○ | ○ | + | | | + | | + | | |
| Low cost | | ○ | + | | | | | | + | |
| Non-contacting | | | | | | + | | | + | |
| In-stream | | | | + | + | | ○ | + | | |
| Density-independent | + | ○ | + | ○ | | + | | | ○ | |
| High penetration | ○ | | | + | + | | + | + | | |
| Ruggedised | | ○ | | + | | | ○ | + | | |
| Intrinsically safe | | ○ | + | | | | | | + | |
| Explosion-proof | | | | | | + | + | | | |
| Ease of calibration | + | | | + | + | + | ○ | | | |
| Ease of application | | ○ | | ○ | ○ | + | ○ | ○ | | |

+, Characteristic met by some models.
○, Characteristic may be met by some models.
Rough preference order (most preferred first): infra-red reflectance; capacitance; microwave attenuation; NMR; equilibrium relative humidity; temperature difference; mechanical properties; neutron moderation; loss in weight.

contents may present problems, and in rare cases spectral interference from a component of the solid may occur.

Cost is complex and requires detailed consideration. Infra-red diffuse reflectance gauges are easily installed in a wide range of plant situations, and require no (or at worst a minimal) sampling system and little complex maintenance. They also have overall a good industrial record for reliability. Hence the total cost, including installation, maintenance and losses due to inefficient operation during sensor down time may be less for infra-red than for an apparently less expensive to purchase sensor.

The complementary argument is that if the measurements cannot justify the installed cost of an infra-red gauge it may not be worth making! However, if the sensor is required before its potential for saving can be found, or if there are a large number of points at which on-line moisture determination is required, there may be real justification for selecting a lower cost method. For dryer control[3,4], the use of temperature differential between that of the drying gas and the exit gas or the relative humidity of the off-gas from the dryer have frequently been used with success. Installation of such systems is generally neither expensive nor difficult, the points to bear in mind being in one case adequate thermal contact between the solids and their temperature sensor, and in the other protection of the moisture sensor from any fine dust carried over from the dryer. Since a dryer is rarely a system in static (as opposed to dynamic) equilibrium, known values for equilibrium relative humidity and temperature depression cannot be assumed, and an empirical correlation has to be obtained between the temperature difference in the off-gas relative humidity and the moisture content of the solids emerging from the dryer.

Where a simple physical property, such as the ability of moist sand to bridge gaps of differing size, is applicable, this also offers a low-cost alternative to infra-red diffuse reflectance.

However, in all these cases the cost of the empirical correlation between the measured property and solid moisture content, and that of possibly poorer dryer control (not selling water or loosing energy) must properly be taken into account in assessing the final cost of the system.

As infra-red diffuse reflectance only 'sees' the surface layer of the material, non-uniform distribution within the grains, lumps or strands being examined can cause false readings. Surprisingly, in many applications where non-uniform distribution is known or is very likely to be present, very successful performance has been achieved, so that suspicion of non-uniform moisture distribution is not a reason for automatically

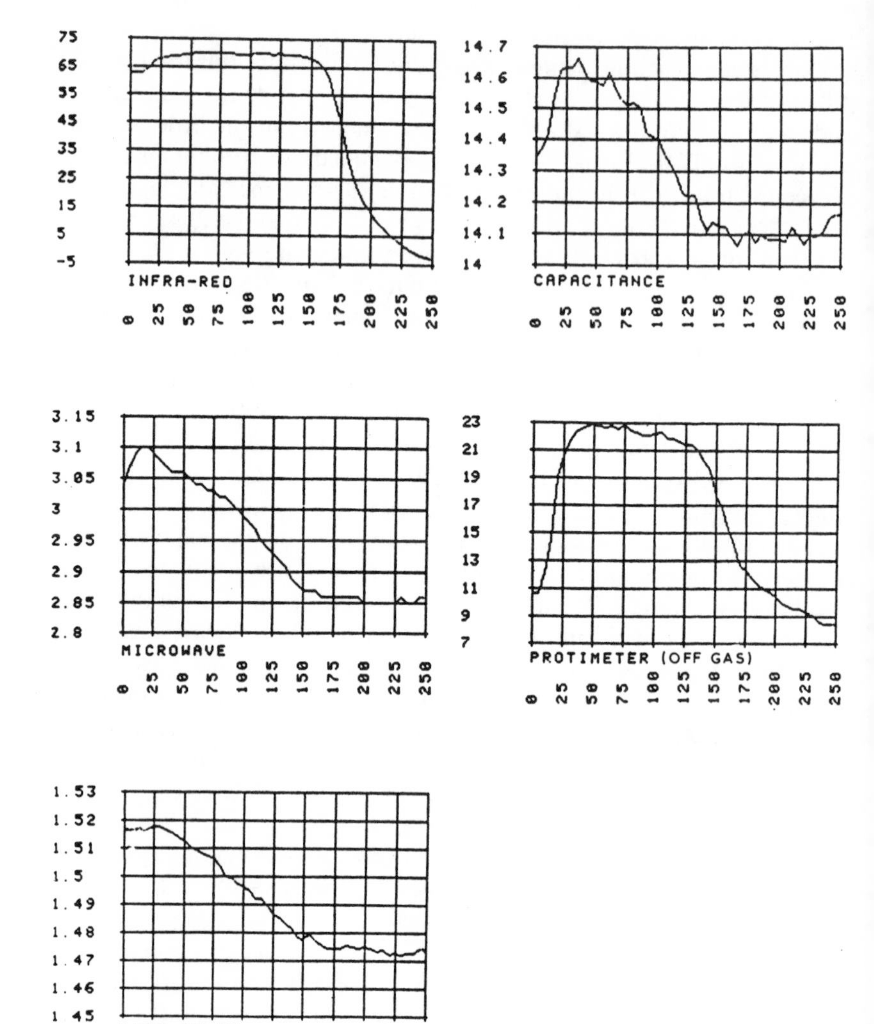


**Fig. 16.** Comparative sensor outputs when drying maize grits.

ruling out infra-red diffuse reflectance. With dryers, non-uniform distribution of moisture with depth in the individual grains or fibres is highly likely; yet this very factor can be used to give high sensor sensitivity near the end point of drying—especially with coarser grained material—which can simplify control strategy (see Fig. 16). Where there are variations in moisture distribution across a strand—such as with material being conveyed on a belt—simple mixing and surface finishing with a rake-like device can obtain very good correlation between infra-red gauge output and water contents measured off-line after repeated representative sampling.

Very high water contents can cause problems with infra-red diffuse reflectance due in part to poorer sensitivities in the higher concentration region, and partly to the ease with which a non-representative surface can be formed. Empirical methods of immediate premixing or the selection of a shorter wavelength moisture absorption band may improve the performance, but other means of moisture determination may have to be sought at such concentrations.

High infra-red absorbance over a broad wavelength band and specular (mirror-like) reflectance can cause problems in the application of infra-red diffuse reflectance—either due to lack of signal or erratic very high 'spikes' or noise as the beam is specularly reflected back into the detector from the correctly aligned surfaces of individual grains or particles. Coal is a material that can exhibit both these problems. The former can be overcome to a certain extent by increasing infra-red source power and increasing the sensitivity of the detector, and the latter by finer grinding, but this may result in infra-red diffuse reflectance not being suitable for such materials. Where the solid material contains significant amounts of coarse material (0.1 to 10 cm or greater) then real problems can arise with non-uniform distribution of moisture and the specular reflection problem can be made worse. In these situations alternative methods of determining the moisture content have to be considered. The most readily available are capacitance (or dielectric constant), microwave absorption, nuclear magnetic resonance and neutron moderation.

Capacitance methods are relatively inexpensive, are available in very rugged housings and have a wide range of head configurations, allowing the method to be used in a corresponding range of plant situations.

However, the method is dependent on the bulk density of the material—this may need independent measurement in cases where it cannot be controlled—and can be interfered with by variations in the concentration of dissolved electrolytes within the material. Nevertheless, by measuring at differing frequencies it can give information on the form in which the water is present—see Chapter 6. Microwave attenuation is less sensitive to the effect of dissolved electrolytes and can be made bulk density-insensitive over quite useful ranges. However, it tends to be more expensive and problems have been encountered with physical factors such as standing waves. Another problem is the limited number of suppliers, especially in the lower price bracket and for the bulk density insensitive models.

Nuclear magnetic resonance has only limited applicability for determining moisture in solids—it being particularly valuable with pumpable material such as high solid content pastes. Neutron moderation has a special niche where deep penetration and extreme ruggedness are essential such as in the iron and steel, non-ferrous metals and cement industries. It does have the disadvantage of employing a radioactive source or sources emitting neutron and $\gamma$ rays, although this can be minimal in industries or on sites where these are already in accepted use. As the neutrons can induce minute levels of radioactivity within the materials indicated, it is not an acceptable method for the food industry—even though the induced levels may be well below the natural radiation in the food itself, arising from for instance, the naturally occurring radioactive potassium isotope. Selection of which of these four techniques to use is often complex, and may be guided by factors such as cost, ease of installation or proven application data by the manufacturers concerned. As a general rule, the order of attractiveness is capacitance, microwave absorption, neutron moderation and nuclear magnetic resonance, with the last being the least attractive for most solid materials.

## REFERENCES

1. Nahar, R. K., Khanna, V. K. and Khokle, W. S., *J. Phys., D. Appl. Phys.*, **17** (1984) 2087–95.
2. Chen-Hsi, Lin and Senturia, S. D., *Sensors and Actuators*, **4**(4) (1983) 497–506.
3. Hebrank, W. H., *J. Eng. for Industry*, **102** (1980) 347–51.
4. Seagar, H., Taskis, C. B. and Way, T. J. R., *Manufacturing Chemist* (February 1976) 31–9.

# CHAPTER 10

# *Sensor Sources*

Process control sensors are normally purchased from commercial manufacturers, who can provide applications expertise, maintenance and 'trouble shooting' as well. However, if no suitable commercial sensor is found, there are various centres of excellence where new sensors can be developed and existing ones modified to meet required needs. However, the cost of such a development is likely to be high (£10 000–50 000) and the time scale in months or even years rather than days or weeks. Hence, there is a strong incentive to either find ways round making the measurement or to use a commercial sensor that will give a 'better than nothing' performance.

In this chapter, the sensors are grouped under the Section numbers given in the previous chapters. After the suppliers name an M indicates a manufacturer and A an agent within the country of their address. At the end is a list of manufacturers' addresses, followed by some UK centres of especial expertise in moisture sensors, who may be able to assist the user in meeting their outstanding needs. A rough indication of price is also given. This is the price for the sensor itself, not the installed cost, which may be 2–3 times greater, depending on the complexity of installation. While every attempt has been made to cover most local suppliers, omissions are bound to occur, and the author and publisher can bear no responsibility for these.

## 10.1 SENSOR SUPPLIERS: GASES

**Mechanical (Hair)—Section 2.3.1**

Typical cost range £300–400
Brown Boveri-Kent M
Jumo M
Pye Unicam/Phillips M

**Wet and Dry Bulb—Psychrometers—Section 2.3.2**

Typical cost range £500–1000
Ancom M
Centemp Ltd M
Jenway M
Lee Engineering A
Telemechanics M
Texel A

**Dew point Sensors—Section 2.3.3**

Typical cost range £2000–5000
Auriema A
EG & G M
Endress & Hauser M
General Eastern M
Lee Engineering A
Protimeter M
Michell Instruments M

**Electrolytic (Phosphorus Pentoxide) Sensors—Section 2.3.4**

Typical cost range £1000–3000
Anacon M
Beckman-RIIC M
Du Pont M
Fisher Control M
Interatom M
Lee Engineering A
Salford Electrical Instruments M

**Lithium Chloride Hygrometers—Section 2.3.5**

Typical cost range £500–1000
Jumo M
Lee Engineering A
Pye Unicam/Philips M
Siemens M

**Aluminium Oxide Hygrometers—Section 2.3.6**

Typical cost range £1000–2000
Endress & Hauser M
General Eastern M
Lee Engineering A

MCM M
Michell Instruments M
Molecular Controls M (also $Ta_2O_5$)
Ondyne M
Panametrics M
Shaw Moisture Meters M

**Silicon Hygrometers—Section 2.3.7**
Typical cost range £1000–3000
MCM M

**Polymer/Ceramic Sensors—Sections 2.3.8 & 2.3.9**
Typical cost range £600–1500
Anacon M
Ancom A
Centronic Sales M
Lee-Integer M
Lee Engineering A
Novasina M
Phys Chem Research M
PP Controls A
Rotronic M
Telemechanics M
Testoterm M
Vaisala M

**Crystal Oscillator Sensor—Section 2.3.10**
Typical cost range £12 000–15 000
Du Pont M

**Infra-red Gaseous Moisture Sensors—Section 2.3.11**
Typical cost range £3000–5000
Anacon M
Anarad M
Analysis Automation M
Analytical Development Co M
Beckman Industrial M
Combustion Engineering M
Foxboro M
Gas Measurement Instruments M
Hartmann & Braun M

Horiba M
Kent Industrial Measurements M
Leeds & Northrop M
Neotronics M
Servomex M
Sieger-TPA M
Smail Sons & Co A

**Other Gaseous Moisture Sensors—Section 2.3.12**

Mass spectrometric and gas chromatographic analysers represent a large market with only marginal interest for moisture determination. Hence a comprehensive list is not appropriate; however, the following gives suggested contacts in the relevant areas.

*Mass spectrometry*
CVC M
Leybold Heraeus M
V.G. Gas Analysis M

*On-line gas chromatography*
Applied Automation M
Beckman Industrial M
Combustion Engineering M
Foxboro M
Servomex M

*Very high moisture*
Serck Glocon A
Yokogawa Hokushi M

## 10.2 SENSOR SUPPLIERS: LIQUIDS

**Infra-red Sensors—Section 3.2.1**

Typical cost range £3000–6000
Anacon M
Foxboro M
Kent Industrial Measurement M
Servomex M
Sieger-TPA M

**Aluminium Oxide Sensors—Section 3.2.2**
See Section 10.1

**Equilibrium Relative Humidity—Section 3.2.3**
See Section 10.4

**Process Refractometers—Section 3.2.5**
Typical cost range £8000–10 000
Anacon M
Moisture Systems M

**Other Physical Property Sensors—Section 3.2.6**
The range is too wide to give comprehensive lists of suppliers. However, the following are known suppliers in the indicated areas.

*Conductivity*
Endress & Hauser
Foxboro
Kent Industrial Measurements
Leeds & Northrop
Pye Unicam
Siemens

*Viscosity*
Alexander Cardew
Clandon Scientific
Contraves
Englemann & Buckham
Eur Control
Fisher Controls
Porpoise Viscometers
Veb MLW
Viscometers (UK) Ltd

*Density*
Krohne
Paar Scientific
Sarasota Automation
Solartron Instruments

*Capacity*
Endress & Hauser

## 10.3 SENSOR SUPPLIERS: SLURRIES, PASTES AND EMULSIONS

These will be found in the liquid and solids supplier sections, since few manufacturers sell specifically into this market.

## 10.4 SENSOR SUPPLIERS: SOLIDS

**Loss in Weight—Section 5.2.1**
No suppliers of 'off the shelf' systems have been located, although Auriema supply an automatic unit made by Photovolt (USA) which could be incorporated into an on-line system. There is also a range of auto taring digital balances, capable of operating through an IEEE interface, which could be similarly incorporated with a feed system, heater and microprocessor read-out.

**Mechanical Properties—Section 5.2.2**
Normally assembled by the user or a sensing consultant with industrialised sensor as required by the particular method.

**Temperature Difference—Section 5.2.3**
There is a wide range of suppliers of temperature measuring equipment, so that selection is inappropriate. Typical cost for a two channel temperature measuring system—£750.

**Microwave Attenuation—Section 5.2.4**
There is a lack of manufacturers of suitable on-line equipment:

Typical cost range £6000–8000.

BFMRA M—made to order, mainly for food and related industries
Kay Ray Inc M—duct mounting with provision for automatic $\gamma$ ray density correction M
Rosemount A

**Capacitance Measurement—Section 5.2.5**
Typical cost range £4000–5000

Brabender Messtechnik M—true on-stream sensor with a variety of heads; available for differing situations
(UK Agent Englemann & Bukham Ancillaries) A
Data Tech M
(UK Agent Auriema) A
Diversfield Engineering Inc M
(UK Agent Laboratory Impex) A
Moisture Register Co M
(UK Agent Shields Instruments (Sales) Ltd) A

**Infra-red Reflectance—Section 5.2.6**
Typical cost range £5000–10 000
Anacon (Instruments Ltd) M
Data Tech M
(UK Agent Auriema Ltd) A
IMS M
(UK Agent Auriema) A
Infra Red Engineering M
Moisture Register Co M
(UK Agent Shields Instruments (Sales) Ltd) A
Moisture Systems Ltd M
Pacific Scientific M

**Low Resolution Nuclear Magnetic Resonance (NMR)—Section 5.2.7**
Typical cost range £15 000–20 000
Oxford Analytical Instruments M

**Neutron Moderation—Section 5.2.8**
Typical cost range £5000–15 000
Kay Ray Inc M
Laboratory Berthold M
Nuclear Enterprises M
Rosemount A
Texas Nuclear M

**Equilibrium Relative Humidity—Section 5.2.9**
This method requires the accurate measurement of the relative humidity of the gas in equilibrium (or in on-line application, often in reproducible contact) with the solid. Hence any of the range of gaseous moisture sensors in Section 10.1 can be used. While the typical cost range is the

same, installation for many samples may report a significant increase in purchasing cost.

**Other Methods—Section 5.2.10**

Analysis Automation Ltd M—automated Karl-Fischer apparatus—requires sample handling system for on-line use

Batch Plant Services M—conductance

## 10.5 MANUFACTURERS' ADDRESSES

These are normally the head office, when within Europe, and agents when a European subsidiary has not been located. Some other local agents are also included.

Anacon (Instruments) Ltd
St Peters Road
Maidenhead
Berkshire SL6 7QA UK
Tel: Maidenhead (0628) 39711
Tx: 847283

Analytical Development Co
Pindar Road
Hoddesdon
Herts EN11 0AQ
UK
Tel: 0992 469638
Tx: 266952

Anacon Inc
F.C. Box 416
South Bedford Street
Burlington
MA 01803
USA
Tel: 617 272 9002
Tx: 951733

Anarad Inc
534 E Ortega Street
Santa Barbara
CA 93103
USA
Tel: 805 963 6583
Tx: 910 334 3474

Analysis Automation Ltd
Southfield House
Eynsham
Oxford OX8 1JD UK
Tel: 0865 881888
Tx: 837509

Ancom Ltd
Devonshire St
Cheltenham
Glos GL50 3LT
UK
Tel: 0242 513861

Auriema Ltd
442 Bath Road
Slough
Berks SL1 6BB
UK
Tel: 06286 4353
Tx: 847155

Batching Plant Services
Bestwood Road
Brookhill Industrial Estate
Pinxton
Nottingham NG16 6NS
UK
Tel: 0733 813181
Tx: 377620

Beckman Industrial
6 Stapledon Road
Orton Southgate
Peterborough
PE2 0TB
UK
Tel: 0733 237055
Tx: 32728

Laboratory Dr Berthold
PO Box 160
D7547 Wildbad 1
Calmbacher St 22
FRG
Tel: 07081 3981
Tx: 724019

Berthold (UK) Ltd
Royal House
28 Sovereign Street
Leeds LS1 4BJ
Tel: 0532 458763

BFMRA
Randalls Road
Leatherhead
Surrey KT22 7RY
UK
Tel: 0372 376761
Tx: 929846

Brabender Messtechnik KG
PO Box 350162
D4100 Duisburg, Kulturstr 51–55
FRG
Tel: 0203 770593
Tx: 855317

Alexander Cardew Ltd
2, 3 & 6 Studio Place
Kinnerton Street
London SW1 8EP
UK
Tel: 01 235 3785/6/7
Tx: 916078

Centemp Ltd
Unit 5
Kirby Works
122–4 Heston Road
Heston TW5 0QU
UK
Tel: 01572 6190

Centronic Sales Ltd
Centronic House
King Henry's Drive
New Addington
Croydon CR9 0BG
UK
Tel: 0689 47021
42121
Tx: 896474

Channel Electronics (Sussex) Ltd
PO Box 58
Seaford
Sussex BN25 3JB
UK
Tel: 0323 894961
Tx: 877825

Clandon Scientific
Lysons Avenue
Ash Vale
Aldershot
Hants GU12 5QR
UK
Tel: 0252 514711
Tx: 858210

Combustion Engineering Inc
Process Analytics
PO Box 831
Lemsburg
WV 24901
USA
Tel: 304 647 4358

Combustion Engineering Ltd
Process Analytics
Gunnels Wood Road
Stevenage
Herts SG1 2EL
UK
Tel: 0438 318811
Tx: 826492

Contraves AG
Schaffhauser-Strasse 580
CH8052
Zurich
Switzerland

Contraves Industrial Products Ltd
Times House
Station Approach
Ruislip HA4 8LH
UK
Tel: 08956 30196
Tx: 935129

CVC Products Inc
525 Lee Road
PO Box 1886
Rochester
NY 14603
USA
Tel: (716) 458 2550
Tx: 978269

Data Tech
3110 West Segerstrom Avenue
Santa Ana
CA 92704-0130
USA
Tel: (714) 546 7160
Tx: 910 595 1570

Du Pont (UK) Ltd
Analytical Instruments Division
Wedgwood Way
Stevenage
Herts SG1 4QN
UK
Tel: 0438 734083
Tx: 825591

Du Pont Company
Analytical Instruments Division
Concord Plaza
McKean Building
Wilmington
DE 19898
USA
Tel: 302 772 5500

Du Pont de Nemours GMH
Postfach 1509
D6350 Bad Nauheim 1
FRG
Tel: 06032 3961
Tx: 0415 547

Du Pont de Nemours SA
Boite Postale 85
F91943 Les Ulis Cedex
France
Tel: 06 907 7872
Tx: 691576

E.G. & G. Environmental Equipment D N
151 Bear Hill Road
Waltham
MA 02154
USA
Tel: 617 890 3710
Tx: 923429

Endress & Hauser GMH
Haupt Strasse 1
D7867 Maulburg
FRG
Tel: 076 22 280
Tx: 07 73226

Endress & Hauser (UK) Ltd
Ledson Road
Manchester M23 9PU
UK
Tel: Manchester (061 998 0321) or 01 866 8891—London Office
Tx: 668501 (Manchester) 8814134 (London)

Englemann & Buckham Ancillaries Ltd
William Curtis House
Alton
Hants GU34 1HH
UK
Tel: 0420 82421
Tx: 858891, 858194

Eur-Control (GB) Ltd
222a Addington Road
Selsdon
Croydon
Surrey CR2 8GH
UK
Tel: 01 651 12267
Tx: 24904

Fisher Controls Ltd
Measurements & Analysis Division
Century Works
Lewisham SE13 7LN
UK
Tel: 01 692 1271
Tx: 22469

Foxboro (GB) Ltd
Redhill
Surrey RH1 2HL
UK
Tel: 0737 65000
Tx: 892852

The Foxboro Co
Foxboro
MA 02035
USA
Tel: 617/543 8750
Tx: 92 7602

Gas Measurement Instruments
Inchinnan Estate
Renfrew PA4 9RG
UK
Tel: 041 812 3211
Tx: 779748

General Eastern Instrument Corp
50 Hunt Street
Watertown
MA 02172
USA
Tel: 617 923 2388
Tx: 710 327 1444

Hartmann & Braun
Moulton Park
Northampton
Northants MN3 1TF
UK
Tel: 0604 46311
Tx: 311056

Horiba Instruments Ltd
5 Harrowden Road
Bracknell
Northampton NN4 0EB
UK
Tel: 0604 65171
Tx: 311867

Humitec
Peter Greaves & Associates Ltd
PO Box 30
Horsham
West Sussex
UK
Tel: 0403 730730

Infra Red Engineering Ltd
Galliford Road
The Causeway
Maldon
Essex CM9 7XD
UK
Tel: 0621 52244
Tx: 995266 Emkay G

Interatom GmbH
POB D5060
Bergisch Gladbach 1
FRG

Jenway Ltd
Gransmore Green
Felsted
Dunmow
Essex
UK
Tel: 0371 820122
Tx: 81776

Jumo Instrument Co Ltd
The Maltings
Station Road
Sawbridgeworth
Herts CM21 9JX
UK
Tel: 0279 725501
Tx: 817820

Kane May Ltd
Burrowfield
Welwyn Garden City
Herts AL7 4TU
UK
Tel: 07073 31051
Tx: 25724 KAMAY G

Kappa Janes Ltd
27 Stewart Avenue
Shepperton
Middlesex TW17 0EW
UK
Tel: 09328 62772

Kay Ray Inc
516W Campus Drive
Arlington Heights
IL 60004
USA
Tel: 312 259 5600
Tx: 281085

Kent Industrial Measurements Ltd (EIL)
Hanworth Lane
Chertsey
Surrey
KT16 9LF
UK
Tel: 09328 62671
Tx: 264022

Kent Industrial Measurements Ltd
Howards Road
Eaton Socon
St Neots
Huntingdon
Cambridgeshire PE19 3EU
UK
Tel: 0480 75321

Krohne Measurement & Control Ltd
Galowhill Road
Brackmills
Northampton NN4 0ES
UK
Tel: 0604 66144
Tx: 311544

Lee Integer Ltd
Douglas House
Queens Square
Corby
Northants NN17 1PL
UK
Tel: 0536 201879
Tx: 341543

Lee Engineering Ltd
Napier House
Bridge Street
Walton on Thames
Surrey KT12 1AP
UK
Tel: Walton on Thames 43124/5/6
Tx: 928475

Leeds & Northrup Ltd
Wharfdale Road
Tyseley
Birmingham B11 2DJ
UK
Tel: 021 706 6171
Tx: 336577 LEENOR G

Leybold Heraeus Ltd
16 Endeavour Way
Dunnsford Road
London SW19 8UH
UK
Tel: 01 947 9744
Tx: 896430

Michell Instruments Ltd
Unit 9
Nuffield Close
Cambridge C44 1SS
UK
Tel: 0223 312427
Tx: 32376

Moisture Controls &
Measurement Ltd
Thorp Arch Trading Estate
Wetherby
Yorks LS23 7BJ
UK
Tel: 0937 843927
Tx: 557654

Moisture Systems Corp
120 South Street
Hopkinton
MA 01748
USA
Tel: 612 435 6881
Tx: 951599

Moisture Systems Ltd
The Old School
Station Road
Cogenhoe
Northants
UK
Tel: 0604 890606
Tx: 312463

Molecular Controls Ltd
30 Park Cross Street
Leeds LS1 2QH
UK
Tel: 0532 440814
Tx: 557712

Neotronics Ltd
Parsonage Road
Takeley
Bishops Stortford
Herts CM22 6PU
UK
Tel: 0279 870182
Tx: 817126

Nova Sina AG
Thurgauerstrasse 50
CH 8050
Zurich
Switzerland
Tel: 01 301 4000

Nuclear Enterprises Ltd
Bath Road
Beenham RG7 5PR
UK
Tel: 073521 2121
Tx: 848475

Ondyne Inc
1090A Shary Circle
Concord CA 94518
USA

Oxford Analytical Instruments Ltd
20 Nuffield Way
Abingdon
Oxon
UK
Tel: 0235 32123
Tx: 83621

Paar Scientific Ltd
594 Kingston Road
Raynes Park
London SW20 8DH
UK
Tel: 01 542 9474
Tx: 945632

Pacific Scientific
2431 Linden Lane
Silver Spring
MD 20910
USA

Pacific Scientific
62 Norden Road
Maidenhead
Berks SL6 4AY
UK
Tel: 0628 34450
Tx: 847287

Panametrics Ltd
2 Justin Manor
341 London Road
Mitcham
Surrey
UK
Tel: 01 640 2252
Tx: 8811645

Panametrics Ltd
Shannon Airport
Shannon
Ireland
Tel: 61377
Tx: 26261

Panametrics SRL
Via C Battisti 7
21100 Varese
Italy
Tel: 0332 242224
Tx: 322384 PANAI

Panametrics BV
Westerdorpsstraat 4
Postbus 111
3870CC Hoevelaken
The Netherlands
Tel: 03495 36444
Tx: 79397

Panametrics GmbH
Jenaer Strasse 3
6200 Wiesbada
FRG
Tel: 061 22/4091
Tx: 4182578

P & E Laboratories Ltd
28 Atheneum Road
Whetstone
London N20 9AE
UK
Tel: 01 445 7683

Phys-Chemical Res. Corp
36 West 20th Street
NY 10011
USA
Tel: 212 924 2070
Tx: 620706

Porpoise Viscometers Ltd
Peel House
Peel Road
West Pimbo
Skelmersdale
Lancs WN8 9PT UK
Tel: 0695 27201
Tx: 628632

PP Controls Ltd
Cross Lances Road
Hounslow TW3 2AD
UK
Tel: 01 572 3331
Tx: 934165 PPCLTD G

Protimeter Ltd
Meter House
Fieldhouse Lane
Marlow
Bucks SL7 1LX
UK
Tel: 06284 72722
Tx: 849305 PMETER G

Pye Unicam Ltd (Philips)
York Street
Cambridge CB1 2PX
UK
Tel: 0223 358866
Tx: 817331

Rosemount Engineering Co Ltd
Heath Place
Bognor Regis
West Sussex PO22 9SH
UK
Tel: 0243 863121
Tx: 86218

Rotronic AG
Badenerstrasse 435
Postfach 8040
Zurich
Switzerland
Tel: 01 492 3211
Tx: 822530

Sarasota Automation
King's Worthy
Winchester
Hants SO23 7QA
UK
Tel: 0962 883200
Tx: 47189

SEI—Salford Electrical Instruments Ltd
Peel Works
Barton Lane
Eccles
Manchester M30 0HL
UK
Tel: 061 789 5081
Tx: 667711

Serck Glocon
St Lukes Street
Southgate Street
Glous GL1 5RE
UK
Tel: 0452 28631
Tx: 43139

Servomex Ltd
Crowborough Sussex
TN6 3DU
UK
Tel: 08926 2181
Tx: 95113

Shaw Moisture Meters
Rawson Road
Westgate
Bradford BD1 3SQ
UK
Tel: 0274 733582
Tx: 51598

Shields Instruments (Sales) Ltd
Wheldrake
York YO4 6NA
UK
Tel: 0904 89329
Tx: 57751

Sieger TPA Ltd
31 Nuffield Estate
Poole
Dorset BH17 1RZ
UK
Tel: 0202 676161

Siemens AG
Bereich Mess Und Projesstechnik
E681
Postfach 211262
D7500 Karlsruhe 21
FRG
Tel: 0721 5954411
Tx: 7825569

Siemens Ltd
Eaton Bank Trading Estate
Varey Road
Congleton
Cheshire CW12 1PH
UK
Tel: 02602 78311
Tx: 8951091

Smail Sons & Co Ltd
129 Whitefield Road
Glasgow G51 2SE
UK
Tel: 041 445 4431
Tx: 777360

Solartron Instruments
Victoria Road
Farnborough
Hants GU14 7PW
UK
Tel: 0252 544433
Tx: 858245

Solomat SA
Ballain Villiers 91160
France
Tel: 6-934 5003
Tx: 690380

Solomat SA
Woodbury
Exeter
Devon
UK
Tel: 0395 32199
Tx: 42497

Sovereign Chemical Industries Ltd
Barrow-in-Furness
Cumbria LA14 4QU
UK
Tel: Barrow 25045 (STD 0229)
Tx: 65261 SOVCHEM

Telemechanics Ltd
St Mary's Works
Krooner Road
Camberley
Surrey GU15 2QR
UK
Tel: 0276 25107
Tx: 858623

Terwin Instruments Ltd
Tollemache Road
Spittlegate Level Ind Estate
Grantham
Lincs NG31 7UN
UK
Tel: 0476 65797
Tx: 377675

Testoterm Ltd
Old Flour Mill
Queen Street
Emsworth
Hants PO10 7BT
UK
Tel: 02434 77222
Tx: 86626

Texas Nuclear
PO Box 9267
Austin
TX 78766
USA
Tel: 512 836 0801
Tx: 77 6413

Texcel Ltd
13 Cunningham Hill Road
St Albans
Herts AL1 5BX
UK
Tel: 0272 54482/68490
Tx: 22333 CBSL G

Toshiba International Co Ltd
Audrey House
Ely Place
London EC1N 6SN
UK
Tel: 01 242 7295
Tx: 265062

Toshiba International
5354 South 129 East Avenue
Tulsa
OK 74134
USA
Tel: 800 331 3377

Vaisala Inc
2 Tower Office Park
Woburn
MA 01801
USA
Tel: 617 933 4500
Tx: 710 348 1332

Vaisala (UK) Ltd
Cambridge Science Park
Milton Road
Cambridge CB4 4BH
UK
Tel: 0223 862112
Tx: 817204

VEB MLW
Prufgeratewerk Medingen
Lesskestrasse 10
Freital
DDR 821

V G Gas Analysis Ltd
Aston Way
Middlewich
Cheshire CW10 0HT
UK
Tel: 060 684 4731
Tx: 668061

Viscometers (UK) Ltd
Newbury House
13(a) Newbury Road
Highams Park
Chingford
London E4 9JH
UK

Yokogawa Hokushin Electric Corporation
9-32 Nakacho 2 Chome
Musashino-Shi
Tokyo 180
Japan
Tel: 0422 54 1111
Tx: 02822 327

## 10.6 CENTRES OF EXPERTISE

National Physical Laboratory
Teddington
Middlesex TW11 0LW
UK
Tel: 01 977 3222
Tx: 262344
(Moisture Standards)

'THESAC'
Warren Spring Laboratory
Gunnels Wood Road
Stevenage
Herts SG1 2BX
UK
Tel: 0438 313388
Tx: 82250
(General Consultancy)

SIRA Ltd
South Hill
Chislehurst
Kent BR7 3EH
UK
Tel: 01 467 2636
Tx: 896649
(Moisture/Gases)

'UMIST'
PO Box 88
Manchester M60 1QD
UK
Tel: 061 236 3311
(Fibre Optics)

AERE Harwell
Harwell
Oxfordshire OX11 0RA
UK
Tel: 0235 24141
(Moisture/Gases)

BFMRA
Randalls Road
Leatherhead
Surrey KT22 7RY
UK
Tel: 0372 376761
Tx: 929846
(Food Industry)

# Index